Understanding Human Diversity

No two people are the same, and no two groups of people are the same. But what kinds of differences are there, and what do they mean? What does our DNA say about race, gender, equality, or ancestry? Drawing on the latest discoveries in anthropology and human genetics, *Understanding Human Diversity* looks at scientific realities and pseudoscientific myths about the patterns of diversity in our species, challenging common misconceptions about genetics, race, and evolution and their role in shaping human life today. By examining nine counterexamples drawn from popular scientific ideas, that is to say, examinations of what we are not, this book leads the reader to an appreciation of what we are. We are hybrids with often inseparable natural and cultural aspects, formed of natural and cultural histories, and evolved from remote ape and recent human ancestors. This book is a must for anyone curious about human genetics, human evolution, and human diversity.

Jonathan Marks has worked in biological anthropology and evolutionary genetics and is presently a professor of anthropology at the University of North Carolina at Charlotte. He has been a visiting research fellow at the ESRC Genomics Institute in Edinburgh, at the Max Planck Institute for the History of Science in Berlin, and at the Notre Dame Institute for Advanced Study in Indiana. His work has been published in *Science* and *Nature*, and his prolific scholarship has appeared in journals ranging from *American Anthropologist* to *Zygon*.

The **Understanding Life** series is for anyone wanting an engaging and concise way into a key biological topic. Offering a multidisciplinary perspective, these accessible guides address common misconceptions and misunderstandings in a thoughtful way to help stimulate debate and encourage a more in-depth understanding. Written by leading thinkers in each field, these books are for anyone wanting an expert overview that will enable clearer thinking on each topic.

Series Editor: Kostas Kampourakis http://kampourakis.com

Published titles:

Understanding Human Diversity

JONATHAN MARKS
University of North Carolina, Charlotte

CAMBRIDGE
UNIVERSITY PRESS

CAMBRIDGE
UNIVERSITY PRESS

Shaftesbury Road, Cambridge CB2 8EA, United Kingdom

One Liberty Plaza, 20th Floor, New York, NY 10006, USA

477 Williamstown Road, Port Melbourne, VIC 3207, Australia

314–321, 3rd Floor, Plot 3, Splendor Forum, Jasola District Centre, New Delhi – 110025, India

103 Penang Road, #05–06/07, Visioncrest Commercial, Singapore 238467

Cambridge University Press is part of Cambridge University Press & Assessment, a department of the University of Cambridge.

We share the University's mission to contribute to society through the pursuit of education, learning and research at the highest international levels of excellence.

www.cambridge.org
Information on this title: www.cambridge.org/9781009534307

DOI: 10.1017/9781009534314

First published 2024

A catalogue record for this publication is available from the British Library.

A Cataloging-in-Publication data record for this book is available from the Library of Congress.

ISBN 978-1-009-53430-7 Paperback

"Appearances may be not only deceptive, but also dangerous. Jon Marks explains with his usual clarity, compassion, and sometimes risqué humor why our diversity both as a species and as individuals is normal, inevitable, and should be celebrated, not condemned. Moreover, 'we are genetically programmed to survive by learning, and to do so in unique, local ways.' What we learn to do, unfortunately, is not always wise or wonderful. Marks shows us how easy it is for our sort of animal to misuse our diversity to excuse selfishness, self-righteousness, the enslavement of others, and even murder. Do yourself and everyone one you know a favor. Read this book."

John Edward Terrell, Field Museum of Natural History, Chicago

"Marks has crafted a beautifully written and amazingly detailed explanation of human diversity – his writing is as detailed and wondrous as the manifestation of this diversity is on our planet."

Rob DeSalle, American Museum of Natural History, New York; author of *Understanding Race*

"There are few people I trust to guide me reliably on the topic of human difference quite as much as Jonathan Marks. *Understanding Human Diversity* is an accessible and wonderfully no-nonsense guide to the state of the current science, as well as the history and ethical dimensions of the perennially fraught question of group difference."

Angela Saini, journalist, and author of *Superior* and *Inferior*

"Over the years, scientists have conjured up analogies and catchy phrases with the aim of making human biology accessible and interesting. Some of the 'stickiest' of these witticisms are actually at odds with the state of the science, and do more harm than good to the popular understanding of human biology. In this book, Jonathan Marks deconstructs nine of these misconceptions in highly entertaining and beautifully written chapters. This latest of Marks's books is a cautionary tale for biologists trying to garner interest in their research, and an enlightening lesson in critical thinking all around."

Leslea J. Hlusko, Spanish National Research Center on Human Evolution (CENIEH), Burgos, Spain

"Jonathan Marks is brilliantly blunt, giving us the perfect antidote to genetic determinism."

Rina Bliss, Associate Professor of Sociology, Rutgers University; author of *Rethinking Intelligence: A Radical New Understanding of Our Human Potential*; *Social by Nature: The Promise and Peril of Sociogenomics*; and *Race Decoded: The Genomic Fight for Social Justice*

For Renée, Peta, and Abby

Contents

Foreword

If you have spent any time in a cosmopolitan city like New York or London you have probably encountered all kinds of different people. You have also probably noticed that they are different in all kinds of ways, not only physically but also culturally. They thus differ in terms of body structure and size, skin color, or hair texture, as well as in the clothes they wear, or the food they eat. I have often heard people confidently say, biologists among them, that they can easily tell a Hispanic from a European, an African from an Asian, and so on. Indeed, there exist both biological and cultural features that are indicative of this. As Jonathan Marks explains in this marvelous book, a blonde person is likely to be from northern Europe, and a dark-skinned person is likely to be from Africa. But he also explains that such differences are average ones; they neither define a person's biological race, nor are they in any way accurate criteria for making distinctions among people. Most importantly, the study of human genetic variation shows that we are all related, being found along a continuum. Of course, if you compare any two people far away across this continuum, they will likely differ in several ways. But this does not entail, as it has been assumed in the past, that there are essential differences among human groups. There is no way that we can accurately divide people into groups that are internally homogeneous and entirely distinct from others, genetically speaking. Most importantly, Marks notes, we need to keep in mind that our evolution is a biocultural one, meaning that our culture and our biology are in constant interaction. Only if we realize this, can we understand human diversity. We are all related in many different ways: we are all family, despite our obvious or less obvious differences. This is what science tells us, Marks concludes, whether some people like it or not.

Kostas Kampourakis, Series Editor

Acknowledgments

This book was certainly assisted by several decades of conversations and interactions with friends, colleagues, and students, in various times and at various places, which would be embarrassing if I mentioned them all by name here. Some of them, however, are: Troy Duster, Jonathan Kahn, Jay Kaufman, Alan Goodman, Dorothy Roberts, Alondra Nelson, Richard Cooper, Karen Rosenberg, Susan Lindee, Rika Kaestle, Adam Johnson, Deborah Bolnick, Evelynn Hammonds, Gregory Starrett, Susan Guise Sheridan, Phyllis Dolhinow, Debra Martin, Nancy Scheper-Hughes, Norman Yoffee, Michael Schiffer, Matt Cartmill, Debra Harry, Veronika Lipphardt, Sarah Franklin, Rob DeSalle, Yolanda Moses, John Jackson, George Hersey, and okay I'll stop there.

Looking back over my career, I owe a lot of people unrepayable debts. I will start with the late great Bill Rathje, who got me into anthropology in the first place. A decade or so later, Che-Kun James Shen got me into molecular evolutionary genetics. I also want to acknowledge the boost that younger me received from the very senior scholars who took an early interest in my work and career, and from whom I informally learned a heck of a lot about evolution, genetics, and human variation: most especially George Gaylord Simpson, Frederick Hulse (from whom I also took my first course in physical anthropology), G. Ledyard Stebbins, and a bit later Ashley Montagu, and later still, Sherwood Washburn. It's still difficult for me to wrap my head around the fact that I actually taught "his" course for a few years. I also thank the senior people I learned primatology from, most especially Sarah Hrdy (who also encouraged my first book on the subject of human variation) and Alison

Richard (with whom I also had the privilege of co-teaching). Even though I only met them a few times, I also owe a great intellectual debt to the work of Stephen Jay Gould, Jonathan Beckwith, and Richard Lewontin.

Several very special people were kind enough to look over some of the material and offer helpful comments and suggestions: Lounès Chikhi, Kostas Kampourakis, Agustín Fuentes, Graciela Cabana, and Karen Strier.

As always, this endeavor would not have been possible without the support of Peta Katz, who gets my heartfelt and deepest appreciation.

1 DNA Is Not Our Deep Inner Core

Meaning and Mendel

Who does not know the most basic fact from the science of genetics, that peas and people reproduce in a similar fashion?

It is taught in high schools. Gregor Mendel discovered the fundamental scientific way that organisms breed, and it works the same way in people as it does in peas. Everyone knows that. They may not remember the specifics, with dominant uppercase A and recessive lowercase a – but they know that humans and peas reproduce basically the same way, because they were taught it, and it's true.

Now I am certainly not going to try and convince you otherwise. But have you ever actually seen peas reproduce? Thanks to the internet, you can readily see videos of plant breeding. The videos of humans breeding, of course, are posted on more restricted internet sites.

There is indeed a legal distinction between watching videos of peas reproducing and watching videos of humans reproducing. Somehow lawmakers and geneticists see the two processes rather differently. Maybe peas and people are less similar than geneticists think; or rather, perhaps the perspective of genetics is not the only meaningful or valid viewpoint.

Actually, human and pea reproduction are only similar from a very specific perspective, that of cells.

Gregor Mendel never thought or said that he had discovered anything about how humans reproduce. The perspective of cells was just beginning to be developed when he published his famous work in the mid-1860s. The idea that there might be a common property in the regeneration of starfish limbs, the degeneration of mutant inbred stocks, and the generation of new organisms had emerged gradually over the latter part of the eighteenth century.

Moreover, the idea that there might be a common property in the inheritance of a particular nose form, the inheritance of a debilitating disease, and the inheritance of the human condition itself emerged over the course of the nineteenth century. All were linked by the cell theory, that life became more deeply and meaningfully unified from the perspective of the cell. The cell theory formed the basis of Samuel Butler's 1878 witticism that "a hen is only an egg's way of making another egg." And the cell theory gave generalized meaning to Gregor Mendel's work on pea hybrids, which is why Mendel's work suddenly became important in 1900, 34 years after he originally published it.

Really thinking about science as a culture, that is to say, approaching it as an ethnographic enterprise, involves unthinking your basic assumptions. In this case, suppose that the perspective of cells, by which human and pea breeding are similar, is a perspective that we reject temporarily; and we rather adopt the perspective of bodies, by which penises, vaginas, erogeny (sensual stimulation), and viviparity (live-birth) are significant characteristics of reproduction in one species but not in the other. That was pretty much the perspective of everyone who ever lived, until mid-nineteenth-century European scientists became the official custodians of such knowledge.

After all, it is certainly not obvious that human reproduction and pea reproduction have anything to do with one another. To put it another way, imagine yourself stranded with a group of people who are passingly familiar with plant husbandry, and know where babies come from – but see no reason to connect them. In fact, they kind of think that anyone who sees similarities between breeding peas and making sweet sweet love must be either stupid or crazy. How would you convince them otherwise?

That is a classic ethnographic issue noted by the anthropologist Bronislaw Malinowski a century ago, as he lived among remote, exotic people who would not acknowledge any relationship between having sex and having babies. A modern scientific explanation that purports to identify an underlying similarity between pea and human reproduction must require a technological infrastructure (at very least microscopes, in this case) to be minimally convincing to an audience of intelligent open-minded skeptics. Even with a microscope, could you convince someone that the pea pollen on one slide and the human sperm sample on the other slide are somehow the same?

The fact that both slides contain male gametes that can unite randomly with female gametes to produce an array of zygotes, which in turn can develop into organisms, is exceedingly counterintuitive, which is why it was never discovered, and presumably would never have been discovered, but for the work of diverse specialists in nineteenth-century Europe.

Inheriting

While we can see the coalescence of otherwise diverse phenomena as crucial to the development of a science of genetics, other entangled ideas conversely needed to be teased apart – notably, inheritances. After all, you "inherit" from your parents the normal and nearly invariant condition of having two arms and two legs, each terminating in five digits. You also "inherit" your complexion, which varies a bit in our species genetically, and is also partly a product of your own exposure to sunlight. Further, you also "inherit" your religion. But you learn your religion, and even though you might well grow to reject it, chances are that you won't, and that this inheritance from your parents will shape your own adult thoughts in subtle yet pervasive ways. And finally, you "inherit" your parents' dining-room set, a material creation subject to legal constraints, like taxation and rival claims.

The first inheritance is genetically "hard-wired" – having four pentadactyl limbs is species-wide and nearly invariant, the result of descent from an ancient fish half a billion years ago with those features. The second "inheritance" is partly genetic, but also significantly a product of the conditions of life, namely exposure to ultraviolet light. A day spent surfing and a day spent studying mammal teeth in a museum basement will have different effects upon the body; so will an ancestry from Nigeria versus one from Siberia. Consequently, complexion develops as far more of a biocultural feature than tetrapod pentadactyly, more variable across our species and less constant over the course of a lifetime. The third "inheritance" is the result of your cultural upbringing, a product of your learning or enculturation. You have even more wiggle room here, because these features are not so much physical or biological as mental and behavioral; and highly diverse over our species. These features are entirely acquired from the external world, yet are generally inherited very faithfully in our species. The microcosm here is language, without which a human is incomplete, but which is learned in childhood in a specific form produced within the history of a specific human population. And the fourth "inheritance" is the physical expression of a legal code that permits the products of human labor, industry, and creativity to be transmitted intergenerationally. In essence the four forms of "inheritance" mark out decreasing correlations between parent and child, due to increasing input from external or "environmental" influences.

With the word "inherit" and the idea of intergenerational transmission embracing such radically diverse properties and processes simultaneously, it is no wonder that human heredity has been the source of so much confusion. It embraces different kinds of processes of intergenerational transmission as well, because we inherit all kinds of things from our parents, in addition to their DNA. The first "inheritance" is in the genes and is the domain of human genetic study; its objects are the parts of the body that are the most strongly canalized (i.e., buffered against genetic and environmental variation, in the terminology of geneticist C. H. Waddington). The second "inheritance" also incorporates genetics, but is studied within a broader field of human adaptability, largely descended from the first studies of immigrants in the early twentieth century. With the gene pool held constant and the environment different, the children of immigrants invariably differ physically in subtle but significant ways from their parents and relatives back in the homeland, for example in stature and head shape. This shows that the particular features of the human body are co-produced by the genetic program and by the conditions of life. These features are more developmentally plastic (having a longish or a roundish head) and less genetically robust (having a head; not having a head is not really an option). Here, what we inherit is a range of physical possibilities, rather than a single uniform feature, whose eventual expression is a result of a negotiation between the genome and the external conditions of life. The third "inheritance" was the major discovery of the earliest professional anthropology in the nineteenth century, namely the "knowledge, belief, art, morals, law, custom, and any other capabilities and habits acquired by [people] as a member of society" – in a word, culture – the stuff that makes us human but isn't in our DNA. And finally, we inherit the things themselves, material culture. Things of value, things of use, things of beauty; evocative things, precious things, old things, made things, bought things, things by which to remember the dead, things through which the dead take care of us – all of these are distinctly human. This is a specifically human form of inheritance. No chimpanzee, after all, ever gave her daughter a twig and said, "Use this well for collecting termites. It belonged to your grandmother."

In other words, the relationship between ancestors and descendants is only minimally revealed by Mendelism. People pass on far more interesting things than merely genes across the generations.

The Human Genome Project

Of the many things we inherit, then, only some are biological, and even those are often strongly influenced by non-genetic factors. Our DNA, our genome, is involved in a complex way in constructing part of what we are.

DNA is a famous molecule, crammed into the nuclei of your cells, with a tiny bit more outside the nucleus, located in your mitochondria (the organelles that generate energy for the cell). DNA is a fairly inert molecule, from which functional molecules are ultimately made. We have learned over the last few decades that DNA is used by the cell to make RNA, which may itself be biologically active, or may in turn be used to make other biologically active molecules, the proteins. RNA is biochemically modified by the cell in various ways to gain its biological activity; likewise, the proteins are biochemically modified before becoming biologically active.

Because of its position at the beginning of the cellular operations, a change in the DNA may be expressed as a change in the RNA or ultimately, as a change in a protein. This change may have large effects, small effects, or no effect at all upon the cell or the body. Like much in science, however, it is often easier to describe what a phenomenon is like than what it actually is or does. In the case of DNA, the set of metaphors that has proven to be of greatest value is from language and information theory. Initially formulated in the 1940s by physicist

Erwin Schrödinger, DNA is regarded as a kind of code, communicating genetic information between the nucleus, where it is stored, and the cytoplasm, where it is decrypted. This genetic information, the DNA, is in turn transcribed, edited, and translated, ultimately being manifested in the physical structures of the body.

So DNA is communicative. DNA is like language. It is also like a blueprint, since it contains the instructions needed to build the cell's proteins. It is also like an atlas, a set of maps directing a fertilized egg through a path of embryogenesis, birth, adolescence, maturity, and senescence. In the world of the literary, DNA is like a lot of things, all derived from Schrödinger's metaphor. In the world of the literal, rather than the literary, all the DNA in a sperm or egg is a genome, and most cells in the body have two genomes. When the Human Genome Project was conceived in the 1980s, its goal was to establish the fine-scale structure of the DNA of a normal genome of a normal human (some observed at the time that this was a Platonic goal, ignoring the reality of individual variation in favor of an imaginary abstraction; it eventually came to be regarded as a reference sequence against which others would be compared). To accomplish this, the Human Genome Project asked for three billion dollars in taxpayer money, which they eventually received. In short measure, the popular science section of bookstores came to be populated by genome-for-the-masses titles invoking codes, blueprints, languages, and maps — alone and in mixed-metaphorical combinations.

The Human Genome Project came to near-completion with considerable fanfare in 2000, with President Bill Clinton declaring, "Today we are learning the language in which God created life." Schrödinger's metaphor is so powerful that there is hardly any other way to think of DNA scientifically, aside from being like human communication in various ways. But that is what DNA is like, which is only related symbolically to what DNA is.

When Is DNA Not DNA?

What, then, does it mean if two decades later Land Rover says, "Adventure. It's in our DNA"? Surely it must be a metaphor of some sort, because automobiles don't actually have DNA. The advertisement is intended to convey something along the lines of, "It's who we really and truly are. No kidding. Deep down inside, our cars are imbued with adventure."

And yet, that isn't really what DNA is or does within your body.

There are indeed some ways in which your status, or identity, or fate could be bound up in your DNA – notably, for diseases and biochemical variants. In most cases, however, DNA diversity is correlated weakly or strongly with physical features, and a phenotypic (i.e., bodily) outcome can at best be probabilistically predicted from the DNA. Why? Because the intervening physiology is complex and only sketchily understood.

Aerdata (a Dutch subsidiary of Boeing) advertises, "Aviation. It's in our DNA." That claim is of particular interest to an anthropologist, since if there is one thing that we can pretty safely say is *not* in our DNA, it is flying. Our bodies are built for walking and running, and maybe a bit of climbing, hanging, and swinging; but flight is not in our DNA. Yet obviously the company is not attempting to make a literal, scientific statement, but rather a metaphorical, neoliberal one.

Or Sony, telling potential customers that high-definition television is in its DNA. The only way that a multinational electronics corporation could have DNA (aside, obviously, from the cells of its employees) is if they are using the term very differently than molecular geneticists do when they envision that famous double-helical sequence of A, G, C, and T nucleotides. DNA here means more than biochemistry, or science. It is a metaphor, and an obviously powerful and evocative one, as different companies are using it so freely. In fact it has become a broader part of contemporary jargon: "It's in my DNA" means "It is a feature deeply embedded within me." But not a feature deeply embedded like having two arms instead of three; rather, a feature deeply embedded like stinginess instead of generosity, which isn't really in anyone's DNA.

"I am cheap. It's in my DNA." And there is no appropriate response to the statement, since it is absurd both scientifically and literally. The DNA is working metaphorically here, identifying stinginess not only as a core property, but as a property so fundamental that it requires no further elaboration.

I am that way, just because – timeless, preformed, changeless – it's in my DNA.

My point is just that nobody was using this metaphor before the Human Genome Project.

In a classic 1995 book, sociologist Dorothy Nelkin and historian Susan Lindee showed how the Human Genome Project was imparting a "mystique" to DNA and transforming it into a "cultural icon," with meanings far beyond the mere biology.

Indeed, although they are rarely part of the formal training of scientists, metaphors constitute an important part of science (See *Understanding Metaphors in the Life Sciences*, in this series, by Andrew Reynolds). Making sense of a phenomenon starts with understanding what it is like. To Copernicus, the earth was like a ball embedded in a sphere revolving around the sun. To William Harvey, the blood circulated around the body like planets circulating through the solar system. To Charles Darwin, nature selects who will survive and reproduce like an animal breeder does, but more subtly. To Thomas Hunt Morgan, the genes are arrayed on chromosomes like beads on a string. To Erwin Schrödinger, the genes are coded instructions, like Morse telegraphy. To Richard Dawkins, genes are selfish, looking out only for themselves, like a paranoid despot in constant peril from all directions.

DNA, Human Nature, and Artichokes

The metaphor of most direct relevance here is the one that imagines our consciousness to be layered, as a bestial inner nature enclosed and suppressed by a civilized outer layer. Like an artichoke, whose tender and tasty heart must be exposed by peeling away the fibrous outer leaves, the human core is imagined to be fundamentally different from its surface. This idea long predates Darwin, although Darwinism gave it renewed vigor as a pseudo-evolutionary narrative. "Darwinian man, though well-behaved," wrote W. S. Gilbert in *Princess Ida* (1884), "at best is only a monkey shav'd."

You may learn airs, this image tells us, but beneath it all you are a beast, an ape, a brute. Without the acquisition of a veneer of civilization, we would all descend into the brutal chaos of the British schoolchildren in William Golding's *Lord of the Flies* (1954). In fact, one of the most well-known expositions of this idea was written centuries earlier by Thomas Hobbes in *Leviathan* (1651), who saw the lives of people without civilization to be "solitary, poore, nasty, brutish, and short."

Acquiring civilized ways can thus tame the animal within, but the animal is always there, ready to be unleashed. By the nineteenth century this had crystallized into an antithesis between nature (inherited, innate) and nurture or culture (acquired, environmental). DNA fits rather neatly into the former slot.

The DNA, the genes, are easily transformed from the cellular biochemicals of science into one pole of a metaphysical antagonism between inherent nature and external culture, at eternal war with each other within the human spirit. That is, of course, scientific rubbish, and the recruitment of DNA into such a model is problematic. Nevertheless, there is a widespread equivalence between "DNA" and "human nature" (in all its complexity), with some scholars even using that as a Darwinian litmus test. The deep core of our being, cultureless humanhood, must be there in the DNA, and must be there as the product of evolution.

That neglects, unfortunately, the major features of human evolution, which undermine the idea that nature and culture are indeed separable, and their effects additive and oppositional. They are not actually like the leaves and heart of an artichoke, but rather more like the eggs and flour in a cake. Some cakes may be too rich or too doughy, but the precise contributions of eggs and flour can't be established because their respective contributions are not additive, and both are necessary.

Consider the two most fundamental evolutionary adaptations of our species: walking and talking. Our genetically based adaptations to nature are unimpeachable: our weight-bearing, rather than grasping, feet; our bowl-shaped pelvis, beneath our center of gravity, rather than behind it; and our spinal column entering the skull from below, rather than from behind. Likewise, the position of our larynx, our small canine teeth, and our oral musculature are genetically based features facilitating speech. Yet we learn to walk, and we learn to talk.

Thus, "genetic" and "learned" are not antonyms. Our most fundamental hard-wired evolved biological adaptations are actively learned by every normal person. As far as can be told from old studies of abandoned children, without a model to imitate, one just doesn't spontaneously walk or talk properly. It is not so much that we have an inner genetic core overlain by a superficial learned patina, but that we are genetically programmed to survive by learning, and to do so in unique, local ways.

And ours is an extraordinarily slow process. It takes a couple of years for us to learn to locomote properly (compared to less than an hour for most other mammals), and far longer to learn to communicate properly – even though if we are genetically programmed to do anything, it is to walk and talk. Once again, it is not that our DNA opposes our cultural existence; it's that our DNA compels us to have a cultural existence.

There is consequently no cultureless state of human existence. It is a contradiction in terms. The idea that there is a deep DNA-based human nature, independent of culture, is false. DNA and culture aren't like that.

Chimpanzees make and use tools, as Jane Goodall famously documented in the 1960s. By a million years ago, our ancestors were using tools successfully in two ways that apes don't: to cut things, and to burn things. Our own more remote ancestors had been cutting things for nearly two million years already, which proved to be such a useful adaptation that the structure of their hands had evolved in concert with their tools. So had the structure of their brains. In other words, culture has been an ultimate cause of our biology for well over a million years. Culture has very deep roots in our evolutionary history; our brains, bodies, and gene pools have been genetically adapting to culture for a long time.

Culture is not just an ultimate cause of the human condition, however. Every human being develops in an environment that determines the expression of their genome. That environment includes naturalistic variables, like the altitude, latitude, and temperature; but it also is strongly cultural. There are more obvious cultural features of the environment, such as diet, language, and religion, but also more subtle features, like labor, stress, and social networks. Thus, culture is both an ultimate, evolutionary component of the human condition, and also a proximate, developmental component of the human condition.

Human nature is cultural in yet a third way. This discussion began a few paragraphs ago with a description of the broad metaphor at play: that the DNA represents a deeper animalistic nature within us, which is cloaked or suppressed by culture (or environment, or nurture), but which is analytically separable. The idea of the cultureless human, the blank slate, has been around for centuries, and is very much itself a cultural idea. That is to say, we conceptualize human nature, our DNA, our inner beast, our *tabula rasa*, and whatever equivalences they may have, in cultural terms. How we regard our own nature is an inherently cultural activity.

There is consequently no separating our DNA from our environment, our nature from our nurture, our biology from our culture. The structure of our hemoglobin may be extractable from our DNA; but the structure of our lives and thoughts is not.

So Is Human DNA Cultureless Nature?

No, of course not.

How about when we read about coalitional male aggression – that is to say, war – "inscribed in the molecular chemistry of our DNA" as some enthusiastic sociobiologists put it? Only in a trivial sense. It is among the many things our DNA permits us to do. Our DNA permits us to gang up and commit violent acts; it doesn't permit us to photosynthesize or to turn invisible.

The fact that humans are able to do certain things, however, doesn't describe human nature, any more than the fact that there are people who can use a manual transmission effectively means that driving a stick shift is inscribed in the molecular chemistry of our DNA. What is inscribed, so to speak, in our DNA is far more generalized: learning to follow abstract rules. It forms the basis of human communication and social interaction. And unlike the rules that govern the behavior of most species, our rules are acquired externally. They are learned; moreover, they are not species-specific, but local and historically produced. And further, as the earliest anthropologists discovered, the rules are commonly nothing short of ridiculous to outsiders.

The important question for any act of human violence, then, is whether it involves following the rules, or not following the rules. Thus, the injunction

"Thou shalt not kill" is understood *not* to apply to a wartime enemy. Killing people is sinful; but if they are enemy combatants, it is heroic. Killing is acceptable in self-defense, but not to effect a change of doctoral thesis supervisor. It may be excusable if the killer is mad (insane), but not if the killer is mad (angry). Or perhaps crimes of passion and crimes of lunacy are equally excusable?

So what does it mean to say that killing is in our DNA, if *not* killing is also there, in our DNA? The crucial variable is the context in which the killing takes place, not the act itself. Is the killing appropriate? And that assessment is very much a cultural variable, a product of history, and instilled by learning. Sure, there are psychopaths who ignore the rules, but they are by definition abnormal.

What we do, what our DNA programs us to do, is to *not* be guided by programming, but by rules of culture that are abstract, invented, learned, interpreted, bent, circumvented, and otherwise constructed and leveraged. It is what we evolved to do, and why our brains are so big that our mothers are imperiled by simply the act of giving birth to us. Human DNA is rendered biologically meaningful culturally. We have evolved genetically to think and to communicate in a zoologically unique way, which involves acquiring a local historically produced variant of human language by listening to it for a few years and learning the arbitrary meanings of specific sounds, words, tones, expressions, and movements, alone and in combinations.

There is obviously a human nature, as distinct from a chimpanzee nature or a dog nature. It consists most fundamentally of walking and talking. But those are not the kind of features that actually enter into arguments about human nature. Arguments about human nature are about behaviors with moral valence, things like sex and greed and violence.

But we were talking about DNA. What has DNA to do with morality?

DNA as a Moral Failing

Consider this statement, from a popular science book of the late twentieth century: "Take first the fact that adult men are slightly bigger than similar-aged women ... A zoologist from Outer Space ... would instantly guess that we belonged to a mildly polygynous species."

What does that mean? First, what does "polygynous" mean? It refers to a social order in which a single male mates with more than one female. The opposite would be "polyandrous." Neither of them is "monogamous," which technically refers to a single marriage partner, but more generally refers to a social/sexual unit consisting of two organisms, no more. The extraterrestrial scientist – who is of course super-smart, but more importantly is entirely objective, and most significantly agrees with the author – is a literary device first employed effectively by Thomas Huxley, to make a scientific proposition sound more compelling by transforming it into a science-fiction proposition. Of course, in reality there are no zoologists in Outer Space, so any speculation on what they might think has far less value scientifically than it may have rhetorically.

The alien scientist is simply a stand-in for the unbiased, wise observer who obviously agrees with you. And what do the two of you agree on? In this case, the proposition that the human species is, naturally and objectively, "mildly polygynous." First off, is there any reason at all to believe it? It is indeed true that a correlation exists among the primates between the extent to which males are larger than females (sexual dimorphism) and their normative social/sexual system. The polygynous baboons and gorillas are on one end of the spectrum, and the monogamous gibbons are on the other. By that primate metric, then, humans are a bit sexually dimorphic in body size, and so you might consider us to be a bit polygynous by virtue of our primate nature. And yet, the social/sexual systems of those same primates also correlate with the sexual dimorphism in their canine teeth. In the monogamous gibbons, the males and females have canine teeth the same size, but in the polygynous species, the canines of the males are much larger than those of the females. And humans have no sexual dimorphism in their canine teeth, like the monogamous gibbons by this metric. And if those patterns of sexual dimorphism reflect different patterns of sexual selection operating in baboons and gibbons, then the human pattern seems to show cross-currents. Moreover, humans have their own patterns of sexual dimorphism that don't parallel anything in our close primate relatives – things like pubic hair, breasts (when not lactating), beards, and overall body composition – and would presumably suggest powerful and non-comparable patterns of sexual selection in our own ancestry.

It looks rather like, in the context of primate biology, our "natural" social/sexual system pretty much sums to zero. It only comes out as "mildly polygynous" if you cherry-pick your primate data and ignore the most prominent features of human sexual dimorphism.

But there is more here than just a poorly structured biological argument. There is also a rationalization in nature for some bad behavior. After all, when the professor catches his wife in bed with another man, he can see that as a crime against the natural order, for he knows that we are *not* naturally mildly polyandrous. But when his wife catches him in bed with another woman, he hopes she will excuse him because he was simply obeying his polygynous biological imperatives; and of course, it's not nice to fool Mother Nature.

In other words, it's not my fault; it's just my DNA! It's human nature, and I'm only human!

"Human nature" in the post-genomic age is a conceptual category to embrace moral behaviors for which you bear diminished responsibility, because you lack completely free will. DNA here stands for loaded dice on the crap table of human life.

A traditional Christian theology holds that you can choose freely between good (God) and evil (Satan). If you choose evil, you'll pay for it with an eternity of unpleasantness. But that presupposes a level playing field. Suppose you are tilted toward evil by virtue of circumstances beyond your control? Then you couldn't really be blamed for going over to the dark side, for the choice wasn't really yours. Contrast, for example, Anakin Skywalker, who chooses the Dark Side freely, against the Pharaoh in Exodus, whose heart was hardened by God Himself! It seems hardly fair, then, to regard Pharaoh in precisely the same way that you might regard Darth Vader. Vader had a choice; Pharaoh didn't.

In light of such theological, mythological, and moral considerations, human nature can quickly become disembodied and reconstituted from human cells and tissues, and become transformed instead into the tilt of the playing field of life itself. Rather than a 50:50 shot at good or evil, suppose you are preformed in some way to lean 80:20 toward an evil act. Then there are now forces beyond your control, DNA forces embedded in your brain cells, which compel you toward the devils and away from the angels. That in turn raises the question of how you can blame someone for an immoral act when they didn't really have an open choice, or voluntary control, in the first place. That is to say, human nature here becomes a denial of free will, which in turn has moral and theological implications.

Or in a secular universe, legal implications. A 2012 study found that when judges believed a criminal act to have been partly caused by biology, they were inclined to sentence more leniently, knocking about 8% off of the imaginary perpetrator's jail time. Whether or not your eternal soul is at risk, a few extra months behind bars is definitely on the table on account of your DNA, or at least on account of what a judge believes about your DNA. DNA has mutated, it seems, into something quite unlike the templates for cellular transcription that biochemists work with. DNA is now the raw material out of which to construct an argument about free will.

This is a strange place for our friend DNA, the double-helical nucleotide polymer, to be.

2 Our Fate Is Not in Our Genes

James Watson, the Nobel laureate, Harvard professor, and head of the Human Genome Project, said to *Time Magazine* in 1989, "We used to think our fate was in our stars. Now we know, in large measure, our fate is in our genes."

He was no doubt trying to find a cute way of saying that scholars used to try and predict the future astrologically, but now science is allowing us to do it genetically. But is genetics really like astrology in that way (although presumably more accurate)? Is the future really predictable from genetics? Of course, there are many genetic diseases known in humans – though relatively few of them are fully penetrant, meaning that if the gene is there, so is the disease. But most features of the human genome aren't like that at all. Rather, the effect of any particular gene depends upon the other genes present, and upon the physiological and environmental contexts in which the gene is expressed.

Fate is a lot bigger than that, however, and hence the cleverness of the sound-bite. Genetics does allow you to tell a person's fate in a small number of cases, most of them tragic. Even though it's not much, it's still presumably better than astrology, and it indeed helped propel the funding for the Human Genome Project through the US Congress.

But let's return to the sound-bite. Is it in fact the case that "in large measure, our fate is in our genes"? If your genes include the one for Huntington's chorea, for which the presence of the gene very precisely predicts the presence of the disease, then you will very likely develop the disease at the end of a life that was nevertheless full of things that genetics can't predict anything at all about.

So, aside from our general appearance and rare crippling or debilitating diseases, what kind of "fate" actually is in our genes? Our waistline? The language we speak at home? Our religious views? The ability to drive a car with a standard transmission? Our maximum running speed? What we find amusing? Our fashion sense? Our vocabulary? Indeed, very little of our lives is genetically fated at all. And even on the genetic enthusiast's best day, plenty of people with the gene for Huntington's chorea have died in battles and plagues long before the gene even had a chance to kill them.

What about our socio-economic class? Is that fated in our genes too?

This is where it gets especially knotty. After all, the bedrock of a hereditary aristocracy is the idea that your parentage determines your status. That particular idea stands in some measure of conflict with the idea that all people are created equal – and this has indeed been a sticking point for democracy over the last few centuries. Is social stratification natural? Is it good? And if so, what should it be based on? And if not, how do we work towards a society of true equality, in which people will be judged by their character and deeds, rather than by the circumstances of their birth?

That is what the molecular biologist jumped into, when he tried to gin up support for the Human Genome Project by declaring that "our fate is in our genes." A lifetime of studying DNA doesn't really prepare you for grappling with the consequences of a public grant proposal that sounds suspiciously like a reactionary political slogan.

Ancestry as Fate

Was James Watson defending hereditary aristocracy? Indeed he was, although presumably without realizing it. "Our fate is in our genes" means many things. To Watson, it meant "Molecular genetics is the most important thing in the world. Please give me three billion dollars." To others who pondered his words, it meant "Ancestry is destiny." And of course, that's cool if you're a member of the nobility; but terribly uncool if you happen to be a peasant or a slave.

Our fate is in our genes. It means, first of all, that we have fates. The life you are living could never have been otherwise, and if it happens to suck, there is nothing to be done about it. It's fated. Or more broadly, it's futile to try and improve your life, because hierarchies are facts of nature, not politics. And as facts of nature, they are not facts of ecology or anatomy; they are facts of genetics. *Our social, political, and economic fate is in our genes.*

So not only do we have fates or destinies, but we also now know that they are located in our genes or ancestries, because ... science! After all, that's what our genes are – the bits of our ancestors that now run our own cells. Our fate is in our genes. Ancestry is destiny.

But do we even have fates, in any reasonable sense of the word? Or perhaps more to the point, if you are, say, non-white – then is your fate in your genes, or in the racism of the people around you, or in the institutions and exploitative relations of earlier generations, or in some noxious combination of all of them? Your genetic status in this case may indeed have a strong effect upon your life course, but primarily in the narrow context in which your society is crudely stratified economically and politically in ways that significantly correlate with genetics. In which case, your fate is in the racism at least as much as it is in your genes. And if we decide that racism is morally evil, then your genes aren't really causing your fate at all; it's the racism. The genes are a red herring.

To target the genes, then, as the significant cause of your life's course, is to imply that racism, sexism, and other forms of injustice are trivial – which is a highly reactionary political position. It certainly had always been so before Watson made his comment. And Watson was, sad to say, hardly the first geneticist to express such thoughtless views.

Even in the late nineteenth century, when cells were being identified as elementary units of life, meiosis was being worked out, and the units of inheritance had not yet been named, a German biologist named August Weismann vigorously promoted his developing theory about the "germ-plasm." Once you recognized that the body is made of cells, and the next generation's body is made up of cells derived specifically from the reproductive cells of the previous generation, you can begin to see life, as we noted last chapter, from the novel perspective of cells. In this case, argued Weismann, studying the history of life involves tracing the continuity of the germ-plasm – the succession of gametes and zygotes over generations. Bodies themselves are merely transient vessels for temporarily storing and transmitting the germ-plasm. More importantly, to Weismann, there was no apparent way for the body to communicate with the germ-plasm and affect it. Mice with their tails cut off for several generation still breed mice with tails. Jewish men with their foreskins cut off for thousands of years still sire Jewish men with foreskins. Thus, argued Weismann, it is impossible to inherit acquired characteristics. Whatever happens during an animal's life does not enter its germ-plasm, and is consequently not transmitted to its offspring. Thus spake cell biology.

The leading British biologist Karl Pearson took that scientific ball in 1892 and ran with it.

> One of the chief features of [Weismann's] theory is the non-inheritance by the offspring of characteristics acquired by the parents in the course of life. Thus good or bad habits acquired by the father or mother in their lifetime are not inherited by their children ... The parents are merely trustees who hand down their commingled stocks to their offspring. From a bad stock can come only bad offspring, and if a member of such a stock is, owing to special training and education, an exception to his family, his offspring will still be born with the old taint. Now this conclusion of Weismann's ... radically affects our judgment on the moral conduct of the individual, and on the duties of the state and society towards their degenerate members. No degenerate and feeble stock will ever be converted into healthy and sound stock by the accumulated effects of education, good laws, and sanitary surroundings.

> If this theory of Weismann's be correct – if the bad man can by the influence of education and surroundings be made good, but the bad stock can never be converted into good stock – then we see how grave a responsibility is cast at the present day upon every citizen ...

Pearson is very generous in attributing his own snooty views about good and bad human "stock" to Weismann, for it is unclear whether Weismann ever derived anything like that from his own work. But the leading biologist in England was using nascent cell biology to make a broad point about the inevitability of social hierarchy. Ancestry, it seems, was destiny, even before genes and DNA.

Freeing the Future from the Past

August Weismann's concern was to argue as forcefully as possible against the inheritance of acquired characters, the fairly intuitive idea that offspring generally inherit their parents' parts, broadly including any quirks, scars, muscles, or memories. Lamarck had made it the engine of his theory of evolution in the early 1800s, but in a broader ethnographic sense, when diverse peoples "are impressed with a spot on the cottontail rabbit's body, they derive it from some adventure of an ancestor of the species, say from a firebrand he once carried in the course of a Promethean exploit," in the words of the American ethnographer Robert Lowie. But August Weismann sought to banish that idea from science. From the perspective of evolution, what happens to you over the course of your life is trivial, since it cannot enter your reproductive cells.

From the perspective of cells, then, your entire life is meaningless, because your life isn't affecting your germ-plasm, which is effectively continuous and immortal. Your germ-plasm is the means by which your ancestors live within you. And that is pretty metaphysical stuff. Might it mean that you can never transcend your ancestors, and the nobility has what they deserve by nature?

Yet while biologist Pearson was invoking cell biology in support of his "neo-aristocratic philosophy," the anthropologist Alfred Kroeber in 1916 invoked the same cell biology in the opposite way. Kroeber observed that since cultural evolution indeed transmits alterations across the generations in response to the conditions of life, the biological isolation of the germ line renders biology more or less irrelevant to human evolution. Who cares if the germ-plasm is not responsive to the environment? Collective human behavior certainly is.

By the 1920s, there were three research strategies available for those scientists who didn't care to feel so completely bound by the germ-plasm. The first was to deny that Weismann had nailed the door shut on Lamarckian inheritance, and to continue to try and find examples by which acquired characteristics were in fact inherited. This first research strategy was dealt two colossal blows over the next few decades, however.

The first blow involved a charismatic Austrian biologist named Paul Kammerer. Kammerer worked with amphibians, which were difficult to breed in captivity, and moreover Kammerer had lost much of his research in World War I. Nevertheless, he continued to promote his work and ideas on the inheritance of acquired traits, and found a receptive audience mired in the pessimism that immediately followed the war. Was there hope at all for civilization after the senseless slaughter of millions? Yes, declared Kammerer. On a 1923 lecture tour in America, he told the New York Times, "Cannot the human race be taught to avoid acquired degenerate tendencies? Cannot the law I have laid down be applied and guide humanity to a higher level? I would suggest, first, that it be used to eliminate race hatred." An admirable thought, to be sure, but not obviously derivable from the reproductive biology of amphibians. Just a few years later, though, it emerged that his prized toad specimen, which had seemingly experienced the inheritance of an acquired trait on its forelimb, had actually experienced an injection of India ink. In September of 1926, Kammerer blew his brains out.

What spark of interest remained in the inheritance of acquired characters was doused in the 1960s with Francis Crick's Central Dogma of Molecular Biology. This was the second blow to the inheritance of acquired traits, the knowledge that the information flow in the cell is one-way. DNA (the genotype) makes proteins in cells (a phenotype), and not vice-versa. In the face of such dogmatism, acquired characteristics would not re-enter the mainstream biological discourse about heredity until the late 1990s as "epigenetics," in reaction to the Human Genome Project. In epigenetics, the focus is not on the DNA nucleotide sequence itself, but on other molecules that interact with the DNA sequence, and which may affect the expression of the DNA, and thus the phenotype, for several generations.

The second research strategy for those scientists interested in softer forms of heredity than Weismann's germ-plasm theory involved following Kroeber's lead and simply studying culture instead of biology. Culture is also contested, having been introduced into English science by E. B. Tylor in 1871 as "that complex whole which includes knowledge, belief, art, morals, law, custom, and any other capabilities and habits acquired by people as a member of society." Obviously as so conceptualized it is specific to humans, and anthropologists have directed their attention to complexities and difficulties with the concept of culture over the years. More recently, animal behaviorists have come to focus on a single aspect of culture, namely its transmission via learning, and have then discovered learned behaviors in other species. However broadly one chooses to define culture, it is fairly clear nevertheless that human behavior is a bit different in being mediated by symbolic vocal language, and in producing sequential change, that is to say, history.

While valuing its distinction from the innate properties of the human organism, anthropologists have also become attuned to the shortcomings of the concept of culture. It can be subdivided endlessly (Western culture? American culture? Southern culture? North Carolina culture?). It is invoked both as a cause and as an effect of human behavior. And it is sufficiently central and multifaceted (like "gene" and "species") that different scholars focus on different aspects of it, making it difficult to define precisely. One studies human culture, consequently, to understand the fundamental technological ways in which human groups adapt to their surroundings, since our ancestors started cutting and burning things over a million years ago. Indeed, even the apparently non-adaptive aspects of culture are what give structure and meaning to human lives. And more importantly, they provide a second window by which people can escape from the tyranny of the germ-plasm. Culture makes descendants different from ancestors in ecologically significant but often non-biological ways.

The third research strategy would be the study of human adaptability. Different human stocks, strains, or races were once considered to be stably different; their distinctions persisted over generations, for they were presumably innate. At the top of the list of stable physical traits of different human stocks was the shape of the head. It could be measured in living and dead people, appeared to be physical and constant through generations, consistently different across populations, and readily quantifiable; moreover, it encased the brain, so it must be especially important. But how stable were heads really? After all, people with weirdly shaped skulls, due to the deliberate modification of infant heads, were known ethnographically – most famously in South America, but all over the world. Might more subtle stresses exist, shaping the head in obscure but consistent ways? It was not until the pioneering work of Franz Boas on the heads and bodies of immigrants in 1910 that the question was systematically examined. Boas measured round-headed "east European Hebrews" and contrasted them with long-headed Sicilians. Not only did their heads start converging in shape in America, but their children's heads converged even more. In other words, the difference in "racial norm" of these groups was disappearing as both groups participated in the hustle and bustle of New York; growing up in urban America had a common, convergent effect on the developing body. Obviously, different peoples weren't turning into one another, but also obviously, at least some of the physical characters that had been thought to be innately and fundamentally different across European populations were in fact not so stable at all, but were subject to extensive input from the environment. To this day, we don't know specifically what features of life in New York affected bodies differently than life in Sicily or Russia did; alas, physiology is complex.

Boas thus pioneered the study of the physical effects of immigration, and, more broadly, of stress upon the growing body, or the study of human adaptability. This is often called developmental plasticity when studied in other species. It acknowledges that the human body is reactive and sensitive to the conditions of growth. That is why identical twins can actually look strikingly different from one another, even though they usually don't. And although we do not know the precise epidemiological pathways by which the life expectancy of a Black person in America is three years lower than that of a White person in America, the study of how a lifetime of struggle or labor or stress inscribes itself upon the human body is now known as embodiment.

Three research strategies consequently emerged over the course of the twentieth century to examine non-genetic continuity and discontinuity over the generations. First, studying the effects of the environment upon the nature of biological heredity itself, from Lamarckism to epigenetics. Second, problematizing the symbolic human environment and its effect upon the survival and adaptation of our species, the diversity of culture. And third, analyzing the biocultural complexity of human growth and development, from immigration to embodiment, the ways in which the circumstances of life are subtly inscribed upon the individual body.

Time and the Genetic Future

Certainly the most unusual feature of the statement that "our fate is in our genes" is the implication that genetics is a deterministic science, when genetics is actually fundamentally probabilistic. That was, in a nutshell, what

Mendel actually discovered. So "our fate is in our genes" actually inverts the principal tenet upon which twentieth-century genetics was based.

What, then, does genetics actually say about the future? Is our fate in our genes, as the scientist James Watson had it? Or is it more like the satirist George Carlin put it, "Everything comes down to luck and genetics. And when you think about it, even your genetics is luck." Genetics in the modern age says there are no good or bad stocks, just a 50:50 chance of receiving any particular allele from a parent.

In which case, what purpose is served by overstating the determinism of genetics upon the life course?

To understand that purpose, we need to recognize the overlaps between theories of heredity and theories of society. For example, the theory of telegony, which holds that a former sexual partner can imprint himself upon a female's future offspring. Its existence in mammals was debated into the twentieth century. In some animals, like fruit flies, the creature's lifespan is not much greater than the sperm cell's lifespan, and after a single copulation a female can fertilize all the eggs she will ever lay. Consequently, her next sexual partner may not be the genitor of any given bubbly baby maggot. For a mammal, it depends upon the lifespan of the sperm cell in a female's reproductive tract (generally a few days) in relation to the arrival of sperm from her next male partner.

If that all sounds fairly arcane, then consider this deduction: Girls who aren't virgins when they marry risk the embarrassment of having a strange baby. Don't be that girl!

And that also explains why anyone outside of prized animal breeders would care at all about telegony. It was really about rationalizing the moral value of female chastity by recourse to imaginary naturalistic consequences.

Indeed, apparently abstract arguments about heredity invariably have social and political implications. There was, after all, a lot at stake in the old days if a son didn't resemble his father particularly closely. Trying to explain those facts would necessarily focus on the behavior of the mother. Was she thinking of another man when her child was conceived? Was it telegony? There had to be an explanation, for her wifely fidelity could hardly be challenged, at least not publicly.

Or, once you know that copulation and semen cause pregnancy, what does the semen actually contribute to the baby? Does it provide the baby's body, which then simply grows in the fertile maternal environment, as the "spermists" had it? Or does the semen activate something inside the woman, who then assumes the burden of making, as well as gestating, the baby, as the "ovists" had it? Does a baby simply expand from microscopic form inside the womb, as the preformationists had it; or does a baby develop from an undifferentiated mass every generation, as the epigenesists had it? And does life differ from non-life in fundamental ways, or can babies come from mud and worms come from decaying fruit?

All of these seemingly biological questions had implications beyond mere biology. If life is continuous with non-life, then maybe life isn't so miraculous. That would certainly be a theological implication; as would its opposite, that a rock is just as miraculous as an elephant. If the baby is really in the sperm, with mom just giving it a place to grow, then that might naturalize gender roles in which men create and women nurture. Would that make a creative woman or a nurturing man unnatural? And if you are the result of the growth of a tiny homunculus embedded within your parent's gonad, and your parent grew from a tiny homunculus within your grandparent's gonad, then you would seem to literally embody all of your descendants. Maybe our fate is in our germ-plasm, after all!

Predestination and Free Will (Again)

And so, once again, we confront our imaginary, inevitable genetic future. The most relevant ancient debate here is between the preformationists and epigenesists of the eighteenth century. Preformism, in which a tiny baby results from the enlargement of a tinier baby, seems to connect the generations more intimately, being contained within one another. Epigenesis, by contrast, in which the embryo is doing more than simply expanding into its appropriate size and shape, seems more amenable to input from other sources than just the ancestors. If we believe in preformism, then, how much of us do we believe to be preformed? Our bodies? Our brains? Our minds? Our moral judgments?

The important question, once again, was not really about the nature of gametes, but rather, to what extent you have the free will to choose good or evil. If you don't choose evil, then you can't really be held responsible for the evil you produce. And if your evil nature is not your fault because it was already there when you were born, then maybe you're not really such a bad person after all. In fact, maybe society owes *you* an apology for trying to make you feel remorseful about your bad deeds.

And that is a lot more important than genetics. Free will seems more harmonious with epigenesis, in which a new person is formed every generation, than with preformism, in which the person (and quite possibly their moral status) is already there. Of course, neither of the eighteenth-century theories is quite right, but they both bear on a central cultural idea: How do we think about the past that we remember and the future that we desire, in relation to the present that we inhabit?

We experience time both cyclically (days, years) and linearly (maturing, decaying). So do our lives have a goal, like a return to the beginning, or a terminal rendezvous at Point Omega, or something else? *Is our fate in our genes? Is our fate anywhere?*

The idea that the future is out there, inscribed somewhere, and we are simply marionettes being manipulated towards that goal, is very old and very pessimistic. As Emperor Palpatine from *Star Wars* might say, "Accept your destiny!" And not only is it pessimistic, for it implies that the future is closed and there is nothing you can do about it, but it is also radically conservative. Since you can't do anything about the course of your future, you might as well not even try. As the Borg from *Star Trek* say, "Resistance is futile." The invocation of fate seems to be all too common among science-fiction villains.

Heroes, like Rey Palpatine (*Star Wars*) and Jean-Luc Picard (*Star Trek*), take control of their destinies. Only ordinary people succumb to fate. And of course, most people are ordinary. The trouble with fate, though, is that there is no way to know whether you are succumbing to it or defying it, because there is no visible single future. Science proves to be pretty good at predicting tomorrow's weather or rare genetic diseases in your family, and pretty bad at predicting tomorrow's winning Lotto numbers or the simultaneous position and momentum of an electron. Generally speaking, science traffics in likeli-hoods, not fates.

Certainly the most pernicious aspect of the statement that our fate is in our genes is the fact that it entirely decontextualizes our genes. Let us say, for example, that you have the genes that allow you to use a solid wooden cylinder to hit a 3-inch sphere coming at you at 95 miles per hour with a fair degree of reliability. Josh Gibson had those genes, but unlike his contempor-ary Babe Ruth he wasn't able to financially exploit his talent for great wealth. Why not? Because Josh Gibson was not allowed to play Major League Baseball, because he was also genetically Black. Josh Gibson's genome did indeed determine his fate, because both his physical abilities and his physical features were genetically strongly "programmed." But it was the context of America in the 1930s that made his talent (a) valuable in the body of a White man, and (b) less valuable in the body of a Black man. And so the course of Josh Gibson's life was determined far more by the popularity of baseball and American racism than by his genome.

What, then, does it mean to focus on the genome, rather than on the racism that dominated his life? Does it mean that you don't really think the racism is that important, and can consequently be downplayed? After all, downplaying the force of racism in American history, and as a determinant of modern life, is a pretty radical position. Some might even call it racist.

Is it racist to say that our fate is in our genes? If not, what is it? Context certainly matters. Watson made the comment as a statement of hereditarianism: Your DNA is the most important thing about you. But hereditarianism does not stand entirely isolated as an "ism." It overlaps other kinds of isms. For an important example, if you live in a society in which the possession of certain genetic markers is a signal that you are only entitled to a lowly place, then presumably your fate is indeed in your genes. In fact, the possession of genetic markers indicating a lowly station in life might well stand as a definition of racism. In this context, then, what does "our fate is in our genes" imply? That you can't do anything to mitigate the racism that structures your life, so you should just suck it up? That sounds pessimistic at best, and enabling at worst. And very quickly, what emerged from the Nobel laureate molecular biolo-gist's mouth as a clever *bon mot* about how important his research was, entered the consciousness of many readers as a reactionary political pronouncement.

Watson maintained that he didn't mean anything politically reactionary, and, as a respected scientist and head of the Human Genome Project at the time, he received the benefit of the doubt. The Human Genome Project got funded (perhaps that was fated!) and Watson continued lecturing, writing, and stumping for molecular genetics. Watson maintained that DNA sequences should not be patentable, which set him against Bernadine Healy, the Director of the National Institutes of Health. He resigned from the Human Genome Project in 1994, and returned to the directorship of the renowned Cold Spring Harbor Laboratories.

But then the sound-bites started re-emerging. Watson publicly suggested that skin pigmentation leads to increased libido, which is why, the honored biologist continued, you have Latin lovers and not English lovers. Har-de-har-har. In a book he published in 2007, he flirted with the idea that Africans were simply innately dumber than Europeans. When he clarified his vulgar ideas on the subject for the British newspapers, his speaking engagements in the UK were cancelled, he was retired from Cold Spring Harbor and made Chancellor Emeritus, recanted his statements and yet nevertheless still swore that he wasn't a racist and never intended any such thing. And finally, in 2019, for a documentary on his life, Watson looked straight at the camera and said that he did, in fact, still believe that Black people were genetically dumber than White people. At which point, he was summarily cut loose from Cold Spring Harbor.

We derive two lessons from this chapter. First, for James Watson, as Maya Angelou once said, "When someone shows you who they are, believe them the first time." And second, more generally, when someone tells you that your life course is genetic, don't waste your cranial energy wondering how true it might be. Wonder instead why an ostensibly smart person is telling you such rubbish. There are no value-neutral statements about human heredity.

3 We Are Not 98% Chimpanzee

It had been well established by the 1920s that human and chimpanzee bloods were more similar to one another than horse and donkey bloods were. This was not particularly unfathomable, since humans and chimpanzees are rather similar, after all, but the special similarity of their blood was well known in educated circles at the time of the Tennessee "Monkey Trial" in 1925.

The similarity of the blood of humans and great apes remained an interesting fact until the 1960s, as the molecularization of biology was proceeding, and genes were shown to be linear instructions for the production of proteins, and proteins were shown to be long chains of amino acids. And when you painstakingly sequenced the 287 amino acids of human hemoglobin and gorilla hemoglobin, showed biochemist Emile Zuckerkandl in 1963, you found them to differ in only one or two places, for a difference of less than 1%. Therefore, continued the biochemist, "from the point of view of hemoglobin structure, it appears that gorilla is just an abnormal human, or man an abnormal gorilla, and the two species form actually one continuous population."

This was a direct challenge to other biologists, who were quite easily able to distinguish a human from a gorilla from pretty much any part of their bodies. It was as if someone had just invented a new kind of telescope that made Venus and the moon look the same. You would probably thank the inventor politely and go about your business, knowing that Venus and the moon are actually very different, even if this new technology does not permit you to detect it. And that was precisely the argument adopted by the paleontologist George Gaylord Simpson, challenging the value of the point of view of hemoglobin structure: "From any point of view other than that properly specified, that is of course nonsense. What the comparison really seems to indicate is that in this case, at least, hemoglobin is a bad choice and has nothing to tell us about affinities, or indeed tells us a lie."

Or more generally, if you can't tell a human from an ape by looking at their genes, just try looking at their feet.

Something Fishy

The feet of a human and a gorilla are made up of more or less the same parts in more or less the same relationships. However, the human foot has been tweaked by evolution to be principally a rigid weight-bearing structure, while the gorilla foot, like other ape feet, is principally a flexible grasping structure. Nevertheless, as feet go in the animal kingdom, that of a gorilla and that of a human are clearly very, very similar.

That similarity, however, is difficult to compare to the similarity of their hemoglobins. Since the protein is a polymer (a long chain made up of simple subunits), you can simply tally up the number of differences, divide by the total number of amino acids in the protein chains, and have a precise accounting of just how different their proteins are – say, 1%. But is that more or less different than their feet are? The dimensionality here is crucial. A protein chain is effectively one-dimensional, and a straight linear comparison may well summarize its similarity to another protein chain. But a foot is space-filling and changes through time, effectively four-dimensional. The feet are neither 0.2% different, nor 2% different, nor 20% different. The comparisons are non-comparable.

So although the intimacy of ape and human blood was known in the 1920s, it first became quantifiable in the 1960s, with various studies of protein similarity; and especially in the 1980s, with DNA sequencing. The data production was complicated and hi-tech, but the analysis was easy: you just add up the differences and divide by the total, and you know how similar the macromolecules of the different species are. In fact, it was almost the reverse of studying anatomy, where you take linear measurements of teeth and bones, and then try to understand the relationships among the complex patterns described by the measurements.

By the 1990s, it was commonplace to hear that we are "just apes" – and specifically, that we are just tweaked chimpanzees, as bestsellers like *The Naked Ape* (1967) and *The Third Chimpanzee* (1992) avowed. They were almost right, but they left out a critical word: "genetically." After all, the word "ape" denotes a particular kind of creature: hairy, living in trees, with long arms, and interlocking canine teeth, a contrast to the human condition. With our wispy body hair and terrestrial habits, we are quite different from that. So to call us apes means that the word "ape" is being used here in a somewhat different sense – as a member of the descent group of apes. We already have a word for that group of species, though, "hominoid." The descent group, or clade, to which we and the apes belong is called the Hominoidea, and it subsumes both apes and humans, because, well, apes and humans are different. That is to say, Hominoidea is "apes and humans," not "apes." Likewise, what might it even mean to call us a "third" chimpanzee, or any kind of chimpanzee, when we are obviously not chimpanzees? You might as well call us fish. Here is why.

A subtle semantic shift had taken place for these apes. Instead of referring to anatomically defined groups, as they traditionally had been, the word was now being used to refer to phylogenetic descent groups. Thus, the "ape" of "naked ape" refers not to hairy, long-armed tree-dwellers, but rather to the cluster of closely related species, the clade, that encompasses those hairy, long-armed tree-dwellers, and which also includes our own species (Figure 3.1). In other words, being an ape no longer depends on

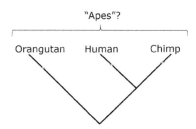

Figure 3.1   Humans fall phylogenetically within the great apes, so on that basis we might regard humans as apes.

what you look like, but only upon your ancestry. If your ancestors were hairy, long-armed tree-dwellers – and ours most certainly were – then you are an ape. Similarly (and rather more controversially), the argument of *The Third Chimpanzee* was that our ancestry places us specifically among the chimpanzees within the apes – which makes us not only apes, but chimpanzees.

The trouble with this reasoning is that whether or not you actually believe we are chimpanzees, when you begin to extend the reasoning more remotely, it quickly becomes very obviously inane.

Consider another anatomically defined creature, the monkey. It has a tail, unlike apes and ourselves, and moves around in the trees quite differently from an ape. Nevertheless, if we take two familiar monkeys, the rhesus monkey and the squirrel monkey, we realize that they are not so closely related to each other. Indeed, the rhesus monkey is more closely related to us (and to the apes) than it is to the squirrel monkey. Like us, an adult rhesus monkey has 32 teeth; the squirrel monkey has 36. Like us, an adult rhesus monkey has a thin septum between its nostrils, which point down; the squirrel monkey has a thick nasal septum, with nostrils pointing out. There are many such features, by which we acknowledge grouping the monkeys of Asia and Africa, the apes, and ourselves, as a cluster of closest relatives, known as the Catarrhini. The monkeys of Central and South America comprise a different, and distant, clade, the Platyrrhini.

But "monkeys" are there in both groups – all of the platyrrhines, and some of the catarrhines. That in turn suggests that monkey life and adaptations are ancestral to the ape life and adaptations (which emerged from a particular form of monkey life), which in turn are ancestral to the human (which emerged from a particular form of ape life).

But at issue here is not what we emerged from, but rather, what we are. So let us apply the same logic to the monkeys as we did to the apes. The rhesus monkey is more closely related to us (as catarrhines) than it is to the squirrel monkey (as a platyrrhine). Therefore, we (and the apes, of course) fall within the descent group delimited by the term "monkeys" (Figure 3.2). And if we define the term phylogenetically, rather than anatomically and ecologically, then we must conclude that we are monkeys.

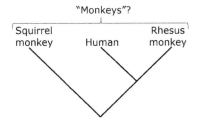

Figure 3.2   But humans also fall phylogenetically within the monkeys. Neither "ape" nor "monkey" refers to a phylogenetic group, or clade. The terms refer to anatomical and ecological relationships, not to phylogenetic relationships.

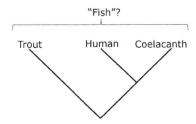

Figure 3.3   By precisely the same phylogenetic criterion that makes us "apes," we are also "monkeys" and "fish." The error lies in mistaking terms that refer to anatomical or ecological relations for terms that refer to phylogenetic relations.

In addition to being apes

Not absurd enough yet? Consider this: The coelacanth is more closely related to us (as tetrapods) than it is to another fish, say, a trout. So if (1) the coelacanth is a fish, which it is; and (2) the trout is a fish, which it is; and (3) "fish" is *not* an anatomical category referring to scaly gilled swimmy stinky things, but a phylogenetic category including all species within that descent group, then (4) by precisely the same argument that makes us apes, we must also be fish (Figure 3.3). Why? Because tetrapods fall within the phylogenetic group subsumed by "fish" – and we are tetrapods.

If, on the other hand, you consider "apes," "monkeys," and "fish" to be anatomical categories, as we traditionally have, then they are different from one another, and from us. They allow us to organize the world in a sensible way – that is to say, these words allow us to indicate that we are not fish, monkeys, or apes. But redefining them as phylogenetic categories which include us makes them useless as classificatory devices.

Some categories are indeed phylogenetic, in addition to being anatomical (at least for living animals), such as "primate" and "mammal." We are indeed primates and we are indeed mammals, by virtue of the anatomical novelties we possess. But apes, monkeys, and fish are different kinds of categories. That is why it is so counterintuitive to hear that we are apes, monkeys, or fish – because whatever meaning those terms have is specifically as a contrast to human.

In other words, we are indeed apes, as long as you define the word "apes" in such a way as to include us.

The Charm of Numbers

When you compare bodies, you compare reactive and responsive living creatures, continually adapting to their circumstances. Bodies are also the products of gazillions of generations of ancestral bodies that successfully replicated and left descendants. When you compare nucleotide sequences, however, you are comparing something else. Now you are comparing things that don't interact directly with the environment, but nevertheless are still the descendants of gazillions of generations of DNA sequences that successfully replicated within bodies. DNA and bodies track the same phylogenetic histories, but in somewhat different ways, requiring different analytical tools. Bodily relations can be obscured by convergent adaptations, while DNA relations can be obscured by convergent mutations.

Since convergent physical adaptations are all but invisible genetically, genetic data tend to reveal the anatomical adaptations of species quite poorly, and phylogenetic relations of species somewhat more readily. That is why the assertion that "we are apes" generally proceeds from the genetics.

The DNA comparisons do more than help reveal phylogenetic relations, however. Since DNA mutations generally do not track the adaptive features of species – again, we don't know precisely how DNA makes bodies, and bodies are what actually interact with environments – the mutations in DNA tend to accumulate in the gene pool in proportion to the rate at which they arise. The rate at which mutations arise is a constant rate. Thus, tabulating the amount of mutations that have arisen in the genomes of two species can permit an estimate of when those gene pools began to diverge from one another. As a result, anatomical comparisons generally track adaptive differentiation of species, and genetic comparisons generally track the time since they diverged.

Even with the crudest genetic measurements, researchers could see by 1970 that the human and chimp genomes were so similar that they could only have diverged from one another a mere handful of millions of years ago – more recently than mouse and rat, or lion and tiger, or horse and donkey. In other words, no matter what our bodies may look like, our genomes are ape genomes.

But of course, they had kind of known that for most of the twentieth century. The crucial innovation toward the end of the century was a change in the meaning of genetic similarity. The genetic similarity went from being an interesting part of the story to being the story itself; the genetic relations became the "real" relations. And thus we went from being genetically apes to being . . . apes.

Ecologically, however, we are very different from apes. Anatomically, we retain the vestiges of ape ancestry, like the short, rigid spinal column and rotating shoulders that allow us to swing through the trees. Except we don't swing through the trees. The apes do. Rather, we walk on the ground, and consequently have bodies that are fundamentally similar to, but distinctly different from, ape bodies.

Similarity is an intuitive, metaphorical concept. That is to say, A is like B in some significant way. We usually don't put numbers on it, but if we did, they would run from 100% similar (i.e., totally the same) to 0% similar (i.e., totally different). Anatomical patterns of similarity began to be teased apart rigorously back in the 1700s. Although one might refer to the "limb" of a person, a horse, and an octopus, the human and horse limbs are clearly much more similar to one another, especially once you get (literally) below the skin. Unlike the octopus, the horse seems to have bones that correspond to our leg bones, ankle bones, and middle-toe bones. That must mean something.

To many Enlightenment biologists, such as Count de Buffon, it meant that God, in His infinite wisdom, "has wished to use an idea, and yet to vary it in all possible ways, so that man could admire equally the simplicity and magnificence of execution of this design." And presumably He used a different design for cephalopods, for some mysterious reason.

And yet the identification of fundamental, underlying similarities among otherwise distinct animals was relativistic and qualitative. You could tell that a monkey's hand was more similar to a human hand than to a horse's hoof or a cat's paw, but just how much more similar it might be was not really an askable question.

That all changed with the comparison of protein sequences in the 1960s, and then with DNA sequences in the 1980s. With these data, not only could you tell precisely how different the species were, but you did not even need a reference point. Thus, when you aligned the sequences of nucleotides (bases) in human and chimpanzee, you generally found them to match perfectly at more than 98% of the nucleotides. At face value, that might indicate that we are genetically about 98% chimpanzee, and that *deep down inside*, we are 98% chimpanzee.

But of course, like the Book of Genesis, you can't really take it at face value.

Genetic Hermeneutics I: The Nature of Genetic Differences

When we encounter the 98% genetic similarity statistic, we are implicitly encouraged to believe that it is (1) a very high number; (2) higher than we might have expected; and (3) a scientific measure of our true relations. Let us take those up in turn.

First, 98% identical sure is close to 100%. That is indicative of near-identity. But how is it measured? One aligns the DNA sequences and tallies up the bases that don't match. The crucial assumption is that base substitutions (and perhaps single-base insertions and deletions) are the only mutations that matter, because they are the only ones that get tallied. They are certainly a big part of the story, but what the last few decades of genomics has shown us is that there is a lot more going on in the genome than simply single-base changes. Not only are there many kinds of mutations besides the single-base changes, or point mutations, but these others are not even included in the count, despite the fact that they may be quite striking.

For example, although point mutations are too small to be seen in an ordinary light microscope, some genetic mutations can indeed be seen. Most notably, if you are looking at a picture of chromosomes and you don't know whether they came from a human or a chimp, all you have to do is count them. A chimpanzee cell has 24 pairs of chromosomes; a human cell has 23 pairs.

Chromosome breakage and reconstitution is a regular mutational mode. In this case, genes that were formerly on separate small chromosomes in the apes have become linked together on what we call human chromosome #2, the second-largest chromosome pair in a human cell. Is it important? Only to those needing help in telling a human cell from a chimp cell. Like most mutations, this one probably has no functional consequences, but it certainly is an obvious one. Somewhere in the human lineage, two different ape chromosomes fused together. There are other structural rearrangements of the chromosomes detectable as well; a few chromosome bits have been turned upside-down in the various lineages. In the gorilla, two chromosomes corresponding to #5 and #17 in the human and chimp have exchanged material, again linking genes in that species that are unlinked in the others.

Many other kinds of mutations in the genome are known but poorly understood. Chunks of DNA can get copied and tandemly inserted next to the original. Repetitive segments of DNA can expand and colonize other regions of the genome. One DNA segment can act as a template to "correct" the DNA copy next to it. Some DNA segments can catalyze their own copying and reintegration at various places in the genome. There is a lot going on in the genome that is simply not captured by point mutations or nucleotide substitutions, and many of them are qualitative changes that are not captured by summing the differences and dividing by the total.

The average 98–99% base-for-base identity between the chimp and human genomes is certainly true and real, but only if you limit yourself to a particularly narrow view of the mutational processes, and only tabulate the things that are amenable to quantitative treatment.

Genetic Hermeneutics II: Are Our Bodies More than 2% Different?

So the 98–99% genetic identity estimate is accurate, but actually very restricted in scope. But is it really counterintuitive? This is where the lack of context for the DNA comparison becomes relevant. Let us say that you sequence a bunch of human and chimp DNA, and find them to be 1.4% different. You say to yourself "Wow, their DNA is 98.6% identical! They are incredibly similar!" Because you are working with an absolute numerical scale, you don't really need an outside reference point, or a context within which to make the statement. You have measured them, and they are 98.6% the same.

But is that actually more similar than their legs? Since legs are not simple polymers, you can't just tally up the differences. You can see that (rather like their genomes overall) the human and chimpanzee body parts are pretty similar, yet diagnosably distinct, if you know what to look for. One is a bit hairier than the other, one is a bit longer than the other, one has a larger gluteus maximus, one has a more flexible ankle. But on the other hand, the two legs are made up of almost perfectly corresponding bones, muscles, nerves, and blood vessels. Compared to the leg of a horse, the basic correspondence is still there, but the horse leg is far more different. And the basic correspondence is still there with the leg of a chicken, but let's face it, no part of that drumstick and your leg are going to be confused for one another.

Nor will anyone confuse your leg for that of a spider or a crab. Those are really, really different. But just how different? They aren't homologous to a vertebrate leg like yours, so is that 0% identical? On the other hand, they are all jointed appendages that support and move us, aren't they? Maybe that ought to be worth something.

The point is that human and chimpanzee legs are quite similar to one another, but the decontextualized comparison is impossible to quantify. DNA comparisons and physical comparisons are hard to compare to one another, at very least because, as already noted, the DNA is a one-dimensional comparison, while the body is a four-dimensional comparison. But even more significantly, once you introduce a context, like the spider's leg, it has no effect on the DNA comparison of human and chimp, but it suddenly makes their legs look very much more alike. In the great panoply of nature, the human and chimpanzee legs match so intimately that they might even be 100% identical, if the frame of reference is a spider's leg.

Consequently, we are left to understand the DNA sequence comparison of, say, 98.6% base-for-base identity as being neither more nor less than we might expect from comparing their bodies. Numerical values are more amenable to one-dimensional comparisons than to four-dimensional comparisons. In fact, back in the 1700s, when chimpanzees were new to science, their astounding physical similarity to humans was indeed what struck observers, long before DNA comparisons. The Swedish naturalist Linnaeus, who never actually saw one, called the chimpanzee *Homo troglodytes nocturnus* – cave-dwelling man of the night. The French naturalist Buffon, who did see a chimpanzee in the 1730s, called it a "Jocko" and doesn't seem to have considered it more than 2% different from a mute human being. Indeed, if we hadn't been studying the differences between the bodies of humans and chimps for 300 years, the measurement of nucleotide near-identity might not have sounded odd at all.

So the measurement of DNA base-for-base identity between human and chimpanzee is hardly counterintuitive or paradoxical or unexpected. It was consistent with what was known about ape and human blood relations, and successfully attached a simple number to something nobody had ever tried to put a simple number on before: precisely how similar we are to an ape.

Genetic Hermeneutics III: The Daffodil Paradox

So humans could be considered over 98% genetically identical to chimpanzees, but only in a very narrow and limited comparison. And we might also be over 98% physically identical to chimpanzees, except that nobody ever thought to summarize it that way before, because it really is quite silly. But maybe we're looking at it all wrong; maybe the DNA base-for-base identity value is simply the best way to compare species, for it provides an unbiased measure of their similarity?

There is no unbiased measure of similarity, for similarity and difference are established culturally. It may not be that human and chimp DNA are so very similar, which places them in adjacent evolutionary boxes; but rather that we don't quite understand the meaning of DNA comparisons, regardless of how close their evolutionary boxes are. Aside from the obvious points that humans and chimps are very closely related and their gene pools only became separated a few million years ago, is the DNA comparison revealing something transcendent about their relations?

We can't answer that question with just the human and chimpanzee genomes.

Let's look, rather, at the bottom end of DNA sequence comparisons – when the sequences in question are very different. "Very similar" or "very different" necessarily imply some sort of scale of comparison. Two things that are 0% identical are completely different. An ice cream cone and a Buick might be 0% identical. A billow of smoke and a dancing bear might be 0% identical, unless of course the billow of smoke seems to look like a dancing bear, in which case you might have a task trying to quantify that resemblance. The Bill of Rights and a plate of linguini might be entirely different, 100% different – although not at the atomic level, where both the paper and the pasta are made of organic plant matter. But what about two DNA sequences? How different actually are two completely different DNA sequences?

We can imagine two DNA sequences that are 100% different, for example a string of As compared to a string of Gs. But that is not our baseline, for that is a contrived, and an entirely unnatural, comparison. You actually have to work to get such a comparison. The appropriate comparison is between two randomly generated sequences, like ACGTCATGCATG and GCATATGCAGTA. And when you compare them, you see that they actually do match in a few places, namely *C*T*****T*. Why do they match a bit? Because there are only four nucleotides in DNA, so any particular nucleotide has a 1 in 4 chance of randomly matching the corresponding nucleotide in the other species. If you compared the sequences of two DNA-based life forms that had no common ancestry at all, they would still be around 25% identical. You could generate DNA sequences that don't match at all, but that would require some tweaking, as we observed above. So the most different that two randomly generated DNA nucleotide sequences could possibly be, without some tinkering, would still be 25% identical. Homologous DNA sequences – those descended from a common ancestral DNA sequence – would obviously be much more similar to one another, hence the 98% or so similarity observed when comparing chimp and human DNA.

What that means is that if you compare yourself genetically to a chimpanzee eating a daffodil, you are over 98% genetically identical in a base-for-base comparison to that chimpanzee, given the limitations noted above. And by exactly the same comparison, you are at least 25% genetically identical to the daffodil the chimpanzee is eating. You are over 25% genetically identical to that daffodil (presumably comparing homologous DNA sequences, such as the cell's "housekeeping genes") because you and the daffodil share a very remote common ancestry. Precisely how similar you are in a base-for-base comparison to a daffodil is hard to establish, because the genomes are so different that it is hard to know which genomic regions are indeed even homologous. But one thing is clear: the DNA similarity between two DNA sequences from different species is statistically expected to be over 25%.

But how would you respond to someone who smugly tells you that, scientifically, you are over one-quarter daffodil?

After all, you are manifestly not one-quarter daffodil. You are only one-quarter daffodil by means of a contrived idiotic comparison, which also happens to be the same comparison by which you are calculated to be over 98% chimpanzee. The obvious conclusion, then, ought to be that DNA comparisons do not afford an unambiguous measure of relatedness – either at the high end, where humans and chimps may seem to be more similar genetically than physically, or at the low end, where humans and daffodils definitely come out to be far more similar to one another genetically than by any more familiar kind of comparison.

The person who tells you that scientifically you are one-quarter daffodil is not doing much to advance science education. It's like saying that that the sky is red or that you are a fish. There are certain conditions that might conceivably make the statements true, but science and everything else say that the sky is blue and that you are not a fish.

Nor are you one-quarter daffodil. No matter what your DNA may say.

Comparing Comparisons

The deduction that we are 98% chimpanzee relies on a lot of cultural facts, most importantly the fact that we have been studying chimps for over 300 years, and we are pretty familiar with how subtly different their bodies are from ours. But we have been studying DNA for only a few decades. That is why the major patterns of DNA comparisons can seem so weird. We simply aren't familiar with DNA comparisons across species to nearly the same extent and depth that we recognize the physical comparisons.

None of the three propositions that went into "We are 98% chimpanzee" is false, but they are all only very narrowly true. You can indeed pluck homologous DNA segments from the genome and find them to generally be 98–99% identical base-for-base. You can indeed decide that that is more similar than their skulls seem to be, because it seems to be obvious. And you can indeed decide that DNA sequence comparisons encapsulate transcendent relations, and insist that you are indeed one-quarter daffodil.

But why bother? Nobody likes a smartass genetics troll.

Ultimately the DNA comparisons reinforce what the anatomy tells us: that we are very closely related to the great apes. But now that we know that, let's take a step back and ask, What purpose would be served by overstating that relationship? As they say in law and bioethics, Who benefits? (*Cui bono?*) What work does "We are 98% chimpanzee" do?

First, by deleting the word "genetically" and thus transforming the statement "we are genetically apes" into "we are apes," we implicitly reduce all relations to genetic relations. The word "genetically" can become invisible once we decide that genetic relations are the only ones that matter. That certainly concedes a great deal of cultural authority to genetics, and of course to geneticists, the scientists who generate, manipulate, and interpret the genetic data. And of course, those are the same data that suggest that you are one-quarter potato.

Second, by focusing on ancestry, which genetic data reveal well, at the expense of adaptive divergence, which genetic data do not reveal well, we reinforce the transcendence of genetics and geneticists over all other forms of biological knowledge and biologists. George Gaylord Simpson recognized this back in the 1960s as molecular evolution was academically rising in ascendancy over paleontology. Of course, molecular evolution and phylogeny are great and very interesting, don't get me wrong. The problem comes when they automatically outweigh any other kind of data or analysis. The point of view of hemoglobin is a legitimate one, but also kind of an eccentric one – but nevertheless a point of view that keeps the DNA sequencers vastly better funded than the bone diggers.

Since we tend to use genetic data as the gold standard in determining ancestry, to focus on ancestry is to privilege genetic data. Schools of scholarly thought that privilege ancestry (notably, phylogenetic systematics, or cladistics) necessarily and synergistically also privilege genetic analyses. But ancestry, or phylogeny, or descent, is only half of evolution. Charles Darwin wrote of evolution as "descent with modification." To focus solely on descent, then, is to regard evolution as effectively descent *without* modification. But both are components of evolution, even if genetics doesn't allow us to identify the bodily modification part particularly well.

Of the 20,000 or so genes in the human genome, only a couple of dozen have been convincingly linked to a regime of selection in a human population. Why is that? First, because humans adapt primarily culturally and physiologically, not genetically. And second, because genes don't directly interact with the environment; bodies do. The adaptations encoded in the genes are there by virtue of mutational changes in our genome that somehow translate physiologically into a better fit to the environment. But we just don't know how to make a four-dimensional body from a one-dimensional set of genetic instructions. Genetics helps us understand ancestry, but if it comes at the expense of devaluing all other evolutionary phenomena, it may be worth questioning those cultural values.

And third, saying that we are apes because we are genetically apes does some rhetorical work as well, against an imaginary creationist interlocutor. It goes something like this: "You think we are specially created? Well, I've got news for you. We aren't just *descended* from apes. We *are* apes! Bwa ha ha!" This line of argumentation works most effectively when one is alone and talking into a mirror, because a real creationist wouldn't be the least bit swayed by it.

After all, if all the evidence that we have ape ancestry doesn't convince you that we are descended from apes, what effect will a geneticist shouting at you that you are an ape have? If a geneticist shouts that they are an ape, you might as well humor them, as if they they're shouting that they are Napoleon.

Actually there are a lot more effective ways of engaging with creationists, especially concerning the humanistic question of how to understand the Bible in the modern age – or in any age. But we can certainly start by agreeing that we are humans and we are similar to, and descended from, apes; and that we fall among them genetically.

4 Human Variation Is Not Race

What Is Race?

The first of several fallacies about race is that it has a precise scientific meaning. Biologists sometimes use the term "race" synonymously with "subspecies" – the lowest taxonomic category, which is created by dividing a species into formal subunits. An example close to home would be the four genetically differentiated populations of *Pan troglodytes*, the chimpanzee. Other biologists reject the subspecies as a taxonomic category altogether.

But whether or not you believe in formally recognized taxonomic human subspecies, as biologists did up into the 1960s, race has always been about the idea that there are a few basic kinds of people. Race is a device for meaningfully sorting people, and we can summarize the idea of race with four properties. First, that there are only a few of them. If there are 700 races of people, then it's not very useful as a classificatory device. Second, that the divisions are made on some natural basis, as unalterable facts of birth: we are not dividing people based on their favorite football team or their language or highest bowling score. Third, the divisions should be fairly discrete, so that most members of the species can be readily sorted or assigned; otherwise it isn't much good for classifying people. And fourth, these divisions presumably ought to map on to significant differences in appearance, temperament, or aptitudes; or else they would be trivial.

Some of these ideas had been around in some form for centuries, reflected in the attitudes of Christians vis-à-vis Jews and Muslims; the English vis-à-vis the Irish; and Europeans vis-à-vis Native Americans. But the ideas crystallized with the booming African slave trade in the late seventeenth century. And for 200 years after the formalization of scientific classification of species in the mid-eighteenth century, if you intended to study humans scientifically, you began by classifying them. And generally you came up with a broadly geographical set of divisions, with perhaps a few famous ethnic groups added, such as the Sami ("Lapps"), the KhoeSan ("Hottentots" and "Bushmen"), the Roma ("Gypsies"), the Irish, and of course the Jews.

What Is Human Variation?

The indistinctness of the concept of race led to long-standing discordances on how many human races existed, what they were, and how they could be delineated. But one thing was unquestioned: They were there. They had to be there. Race would eventually become the classic example of scientists finding the patterns they were looking for, regardless of whether or not the phenomenon was real. Humans, it turns out, are very good at creating and recognizing patterns where none exist in nature, such as constellations or national boundaries. Psychologically, we are especially good at projecting faces onto things, from the surface of Mars to a tortilla. This phenomenon is known as pareidolia, and it is very similar to what scientists were doing with race from the mid-1700s through the mid-1900s.

Early field work in the late nineteenth century began to identify problems with the classic conception of race. For one thing, races were supposed to be fairly homogeneous, yet the closer one examined, say, the "European" race, the more heterogeneity one found. Crudely put, Scandinavians and Iberians tend to look a bit different from one another, as northern and southern Europeans, respectively. By 1899, Europeans were being formally subdivided into Mediterranean, Alpine, and Teutonic or Nordic. By 1939, there were close to a dozen races within the European race. Likewise on other continents, scholars studying actual patterns of human variation found far more variation within any presumptive race than between them.

By the 1960s, students of human variation were finding that the best way to describe its major features was as a series of geographical gradients, or clines. Further, the obvious, interesting, or adaptive features of human variation did not respect continental boundaries. Sickle-cell anemia, for a classic example, is prevalent not in Africa, but rather in those parts of Africa, Europe, and Asia where malaria has been a stressor on the gene pool. "Race" has nothing to do with it.

Geneticists were no less befuddled by the constraints of race. The earliest genetic marker, affording a glimpse of the human gene pool itself, was the ABO blood group system. It was first studied as a genetic marker during World War I, but did not sort out at all racially. Nevertheless, it was interpreted racially, in various ways, into the 1960s. In fact, however, the frequencies of the three major alleles vary only within fairly narrow limits. Consequently, distantly related peoples may have similar ABO allele frequencies purely by chance. Thus, when geneticists discovered that the people of Poland and the people of China had similar ABO frequencies, they did not conclude that the genetic system does not reveal race (as we do now), but rather, that the two peoples comprised an imaginary Polish-Chinese race.

Throughout the middle of the twentieth century, scientists argued about whether genetics was in fact even useful for studying human variation, if it didn't reveal the fundamental human races, which of course had to be there.

The critical question was whether the ABO blood group system was the exception in not being racial, or the rule. A classic study in 1972 by the geneticist Richard Lewontin showed that it was indeed the rule. No matter what genetic system you looked at, chances are that only a small fraction of the total variation is found between any two groups. Overwhelmingly the differences between populations are just slightly different average frequencies of the same set of alleles.

But if race isn't there in the human species or its gene pool, then what is there? Empirically, the patterns of human variation can be studied without presupposing race, and when they are studied like that, it actually becomes very hard to find race at all.

If we wish to formalize the major patterns of variation in our species, the first thing that strikes us, as it struck the earliest nineteenth-century anthropologists, is the "knowledge, belief, art, morals, law, custom," etc., that was called "culture" by the British anthropologist E. B. Tylor in 1871. While human culture (like everything else human) evolved from an ape substrate, nevertheless apes don't seem to possess most of the things on that list in any easily recognizable sense. Obviously apes have knowledge of their environment and social relations, but their knowledge is quite limited, if only for the simple reason that, unlike humans, apes don't ask questions. Likewise, you can define belief in such a way as to include apes, if you believe that apes believe. But even so, what would apes believe in? Polytheism? Patriotism? Ape not kill ape? Free bananas for all? And of course, there are many examples of apes painting, when provided with the appropriate tools for the job. On the other hand, as one might ask with a condescending French accent, "But is it art?" And the apes don't even get that reference.

So, while continuous with the behavioral variation in the apes, humans do something rather different with culture. In particular, peoples use culture to augment and exaggerate biological differentiation. We do it vocally, with our local patterns of speech; and physically, for example, with hair styles. Apes have neither long head hair nor the means to tend it; humans have both the means and the need to do so. Human hair is, as the Broadway song goes, "long, straight, curly, fuzzy, snaggy, shaggy, ratsy, matsy, oily, greasy, fleecy, shining, gleaming, streaming, flaxen, waxen, knotted, polka-dotted, twisted, beaded, braided, powdered, flowered, and confettied, bangled, tangled, spangled, and spaghettied." Not to mention, covered up fashionably. Those simply aren't options for the apes. In our species, hair color and texture vary a bit, but what varies a lot is what people *do* with their hair. Hair is a very powerful and universal visual signal of group identity, of specific places and times. Likewise with other surface features of the human body, like scars, tattoos, and clothes.

Unlike the apes, then, the primary way that humans differentiate themselves from other groups is culturally. Culture is the first thing you see when you meet someone new: how they're dressed, how they speak, how they are groomed – corporeal social information. But suppose, as a scientific student of human diversity, you decided to ignore the major pattern of human variation – the cultural, whose attributes are acquired by learning over the course of one's life – and insisted, however perversely, to examine only physical, natural, biological differences. What would you find? What you would find, as early twentieth-century human biologists found, is that your ambition to separate and study only the physical, biological differences actually fails to distinguish those from the cultural differences.

The organism is invariably a product of the genetic instructions being carried out in a particular environmental context. And of course, "environmental context" is a bit more complicated in people than it is in flies, since it incorporates "knowledge, belief, art, morals, law, custom" and all those other cultural things. Consequently, a lot more affects the growth and development of our bodies than just altitude and sunlight.

Cultural differences become inscribed upon the body in various ways, presumably by epigenetic mechanisms. For example, labor (as the stress on bones induced by repetitive motion). Or class, or gross inequality more generally (as the average differences in stature and health detectable among people of the same time and place). And many other, more subtle features by which, for example, Japanese immigrants to Hawaii, and their children, would be statistically different-looking than their family members who grew up and remained in Japan.

The study of the bodies of immigrants became a critical piece of the scientific study of race in the early twentieth century, when it became clear from contrasting the bodies of transplanted peoples against the bodies of closely related non-migrants, that there was something crucial about where you grow up in determining what you will look like. Immigrants don't look precisely like the kin they left behind. This is not to say that there are no inherent differences in physical form among the peoples of the world, which would be silly. This is simply the discovery that observing a consistent difference among peoples isn't valid evidence for inferring a genetic, innate cause for that difference. Sometimes, like "racial odors," they may be due to labor conditions, or diet, or hygiene, or simply slander. Or perhaps, like the "Dinaric" skull form, the diagnostic differences are due to cultural practices.

Cultural differences often cannot be readily separated from biological differences because any particular physical phenotype is the result of expression of genes in a particular environment. Consequently, similar environmental causes can lead to similar phenotypic effects. Diverse peoples living at high altitudes may have convergent phenotypes. Diverse peoples living in deserts, or tundras, or any climate leading to a stressful existence, may respond physically in similar ways to the stress. And of course, if your interest lies in the study of presumptively racial distinctiveness, the subtler the feature you examine, the more likely you are to discover that it isn't even hard-wired. It has a complex biocultural basis, for ordinarily human growth and development often does not permit the facile distinction of cultural from genetic etiology.

So now let's say that you decide to ignore the major pattern of human variation, the cultural, and you decide as well to ignore the second major pattern of human variation, the developmental. You decide to focus only on variation that is hard-wired, variation that directly translates from genotype to phenotype. Indeed, maybe you can just look at the DNA or genome itself, and forget about phenotypes altogether. What is the major pattern that you discover in the human gene pool?

The pattern you see is that different human gene pools are not very distinct from one another, and that you have to look very hard to find any genetic differences at all. But once you find some genetic differences and focus on them very, very carefully, you see that human populations overlap widely in their genetic variation. It's not that one population only has allele A1 and another population only has allele A2, but rather that one population has 40% of A1 and 60% of A2 and another has 55% of A1 and 45% of A2. In that case, a random A2 allele could have been drawn from either population. Moreover, the farther away the two populations are, the more different they will be from each other, although still not very different. And If we expand our contrast to the human populations on each continent, we find that they don't even make a valid contrast, because (with minor exceptions) we find that the gene pool of African peoples subsumes those of the other continents. This is probably a result of the fact that the earliest people with foreheads and chins, like us, were in Africa, and after a few hundred thousand years, we are all their descendants.

Human gene pools are far more notable for the alleles they have in common than for the ones they don't. This cosmopolitan pattern with nearly ubiquitous genetic variation, or polymorphism, is not what a geneticist of, say, 1960, would have expected to find. But by the 1990s it was clear that it was indeed the underlying pattern of the human gene pool, whether you looked at protein variation, mitochondrial DNA variation, or nuclear chromosomal DNA variation. They all showed that there is about seven times more variation within groups than there is between groups. Consequently, to focus only upon the genetic differences between populations is to miss the major feature of the human gene pool, the extent to which human populations actually overlap genetically.

But let's say, for some strange reason, you are determined to overlook the cultural differences, the developmental physical differences, and the polymorphic or cosmopolitan genetic variation, and focus only upon the tiny bit of genetic variation that differs from group to group. What pattern do you find?

The pattern that you see is clinal, first named in 1938 by the biologist Julian Huxley and identified as the predominant pattern of between-group variation in the human gene pool by the anthropologist Frank Livingstone in 1962. A cline, as noted earlier, is a gradual change in a biological feature over a (usually geographical) gradient, and it represents the major feature of genetic diversity across human populations. The clinal pattern is particularly amenable to being presented in a series of genetic pie-charts, which paradoxically combines some of the highest-tech data with some of the lowest-tech presentation formats.

Undaunted, you finally decide to ignore the primary pattern of human diversity, the cultural; the secondary pattern, the developmental; the tertiary pattern, the polymorphic; and the quaternary pattern, the clinal. What is left?

The pattern of human diversity that remains is the genetic variation from place to place that tracks environmental features adaptively (through the action of natural selection), and tracks demographic histories non-adaptively (through the action of genetic drift). This is the genetic variation that subsumes the adaptations to malaria in the gene pools of early tropical agricultural communities; or the adaptations of Tibetans to the thin air of the Himalayas; or the adaptations to dietary cow's milk in the gene pools of dairying peoples. This is also the genetic variation that encompasses the elevated distribution of the variegated porphyria allele in South Africans of Dutch ancestry, and polydactyly in the Pennsylvania Amish. This pattern is local – perhaps small like the Pennsylvania Amish, or perhaps large like the "Old World malaria belt" – but it certainly doesn't respect continental boundaries, or track anything familiar as race.

Thus, we have five nested patterns of human diversity, evident empirically, and identified by anthropologists over the course of the twentieth century. But race is not one of them; indeed, even thinking about race in the context of human variation simply serves to obscure its actual patterns. Human variation and race are quite different things. To the extent that there remains an interesting unanswered question about the relationship between race and human variation, it lies in the domain of the history and sociology of science.

If race and human variation are indeed so different, then why did scientists think they were the same for 200 years? Science is self-correcting, and indeed corrected itself, yet it took two centuries, and a great deal of social evolution, to recognize that error and to make the correction. How and why?

Where Did Race Come From?

Classifying is as human as walking. How we classify one another was actually the very first research question undertaken by early modern anthropologists, particularly because it seemed strangely counterintuitive that different peoples would conceptualize "relative" or "family" or "cousin" or "father" differently from the obvious English meanings of those terms. Nevertheless, all peoples impose sense upon their social relations, and they do so through the medium of language.

Most importantly, these linguistic categories of kin may be quite tangential to genetic relationships. An "aunt" may be a genetic relative (mother's sister) or a non-genetic relative (mother's brother's wife). And although cousins are equivalent as far as geneticists are concerned, they are not equivalent in the eyes of many human societies. A "cousin" who is your mother's *brother's* child may be considered a very different relation than a "cousin" who is your mother's *sister's* child. To an anthropologist, the former is a "cross" cousin and the latter is a "parallel" cousin; and the difference between them may be the difference between what the people regard as a blessed union and as a sinful, incestuous union – regardless of the fact that they are genetically equivalent.

One common way of imposing order on social relations is as nested circles. I might favor the interests of Charlotte over those of another city in North Carolina, which might be competing for the same pool of resources. Likewise,

I might favor the interests of North Carolina over those of Kansas, and those of the USA over those of Cambodia. And while that may seem a reasonable and natural way to conceptualize social relations and structure one's behavior, we need to bear in mind that competing for a pool of resources (rather than sharing them equitably) is a specifically contrived relationship; and the municipalities, states, and nations ostensibly in competition are units of political history, not of nature.

With the rise of urbanism a few thousand years ago, a new kind of identity was produced. One could now be known from a place of origin, rather than strictly from a lineage. In other words, you might no longer simply be "son of . . . " but in addition (or alternatively) "from . . . " It was also not uncommon to mythologize the origins of the people of a particular place. We see this very early in the Bible, where Egypt is called "Mizraim" and the Egyptians are regarded as descendants of an ancestor called Mizraim (grandson of the ark-builder, Noah). Likewise, the "Canaanites" not only lived in Canaan, but were descended from Canaan (Mizraim's brother, as it happens). So the idea of people from a particular place sharing a special fictive ancestry, as a manner of identification, is an ancient one.

Similarly, the ancients knew of a northern landmass (Europe), a southern landmass (Africa) an eastern landmass (Asia), and the sea in the middle of the lands ("Mediterranean"). Of course, they had no conception of the extent of those landmasses, much less the diversity of peoples upon them. However, by Hellenistic times, the known geographic distribution of people was being understood in the biblical context of the three sons of Noah: Ham, Shem, and Japheth. Ham was the ancestor of the southern peoples, Japheth the ancestor of the northern peoples, and Shem was the ancestor of the eastern peoples (among whom the Hebrews counted themselves, through Shem's great-grandson Eber).

While the continental populations could be crudely conceptualized as descent groups, they couldn't really have been very different from one another, since Ham, Shem, and Japheth were all brothers. More importantly, the differences among various peoples were considered to be local. The people in one place had their own distinctions of speech, religion, appearance, or cuisine; and the people in another place had their own distinctions. The Bible tells us that the Ephraimites spoke a dialect that did not possess the "sh" sound (like Greek), and the Anakites were very tall (like the Dutch). Herodotus (ca. 450 BCE) wrote of the one-eyed Arimaspians and the Argippaei, who were peaceful, bald, and lived on cherry juice – not to mention the Indians with dark skin and dark semen. Pliny the Elder (ca. 70 CE) repeated what he had heard about the Monocoli, "who have only one leg and hop with amazing speed," and the headless Blemmyae in Africa, as well as the Chinese, who traded in fabrics, had golden hair and blue eyes, and made excellent iron. The idea of some sort of basic continental homogeneity would have seemed absurd to them with such extensive diversity locally. And if you took a long trip, you went mostly over land, by caravan, and were struck by the local flavor, and also by the continuity – real people indeed tending to resemble their real neighbors.

But much of that had changed in Europe by the mid-1600s. Now, if you took a long voyage, you went by ship, and you were struck by the discontinuity in human form. Moreover, one of the most powerful reasons for taking a long voyage by ship was to kidnap and enslave people in West Africa, to be sold for profit in the Americas. In that racket, acknowledging distinctions among peoples was positively counterproductive, for scrambling up peoples helped prevent them from banding together against you. Thus, "Africans" became homogeneous by convenience, and distinct from "Europeans" – in spite of Europe's own internal distinctions of nationality, religion, appearance, language, and customs.

Moreover, the discovery of America rendered the origin story of the three sons of Noah for the peopling of the continents obsolete. Perhaps they could have rewritten the Genesis narrative and given Noah a fourth son. Instead, however, mapmakers developed a new metaphor. Rather than being associated with a biblical son of Noah, each continent would be symbolized by the image of a beautiful woman. In Cesare Ripa's illustrated *Iconology* (1603), a guide to symbolic illustration (Figure 4.1), Europe wears a crown, holds a church, and is associated with the trappings of wealth and power, and a horse (this will frequently be replaced by a bull, symbolizing the form taken by Zeus whilst raping Europa); Asia wears a laurel wreath, holds incense (associated with the Eastern church), and stands beside a camel; Africa sports an elephant hat and stands in front of a lion; and America wears a feather headdress, and

(a)

E V R O P A. Vna delle parti principali del Mondo.

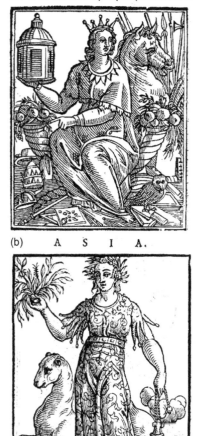

(b)   A   S   I   A.

Figure 4.1   The continents personified, from the second illustrated edition of Cesare Ripa's *Iconology* (1618). Although Ripa specified in the text that Africa should have a dark complexion and frizzy hair, she was usually not drawn that way; and none of the other continents have any physical specifications.

(c)
A   M   E   R   I   C   A.

(d)
A   F   R   I   C   A

Figure 4.1   (Cont.)

brandishes a bow and arrow, standing in front of an alligator. The landmasses, we might say, became symbolically transformed into land-lasses.

Significantly these figures, and their subsequent adoptions by seventeenth-century mapmakers, are unraced; that is to say, they are all facially versions of familiar female Europe. Over the course of the 1600s, however, these illustrations will come to serve double duty – symbolizing not simply the landmasses themselves, as they were initially intended, but also the people on the landmasses. Over the course of the seventeenth century, aided by illustrations like these, the people on each continent became more homogeneous, and became more differentiated from those of the other continents. By 1700, it was common for cartographers to distinguish the four landmasses/women as being not only associated with different items and fauna, but as appearing slightly physically different from one another, particularly Africa.

The symbolic fourfold division of the world, and the symbolic fourfold division of its peoples, was literalized as science by the developer of modern scientific classification, Carl Linnaeus. In his *System of Nature* (10th edition, 1758) Linnaeus formally divided *Homo sapiens* into four geographical varieties or subspecies, differing in color, personality traits, and associations. A century earlier, Europe, Asia, America, and Africa were depicted metaphorically wearing a crown, laurel wreath, feather headdress, or elephant hat; and with a bull, camel, alligator, or lion. Linnaeus now would transform the four landmasses and their women and associations into science. That scientific fourfold division of the peoples of Europe, Asia, America, and Africa would be inscribed as official colors (White, Yellow, Red, Black); personalities based on the dominant Hippocratic humors (Sanguine, Melancholy, Choleric, Phlegmatic); legal system (Laws, Opinions, Customs, Whims); and even mode of attire (Tight-fitting clothes, Loose robes, Painting oneself with fine red lines, and Smearing one's body with grease).

Linnaeus was clearly not discovering that there were four kinds of people in the world. He was transferring a cartographic allegory into scientific data, and shoehorning those data to fit the feet of race.

By the end of the eighteenth century, the leading empirical student of human variation was Johann Friedrich Blumenbach of Göttingen, and he was faced with a dilemma. On the one hand, he recognized the continuity of human form, and the fact that you can't draw a line between large continental groups of people. On the other hand, he was working in the shadow of Linnaeus, the great classifier, and consequently he had to begin with a classification. So Blumenbach tweaked the continental groups that Linnaeus had named – most famously turning "*Homo sapiens europeaeus*" into "Caucasians" – and added the newly encountered Oceanic peoples.

Concurrently, in late eighteenth-century France, the Count de Buffon was surveying human diversity, but not in the classificatory framework that Linnaeus had constructed in Sweden. Buffon, rather, presented human diversity in travelogue form as the ancients had, emphasizing continuity and local diversity in human appearance and behavior. Moreover, Buffon did not make his human groups into any sort of formal subspecies; he thought of them as descent groups, like strains or breeds of animals. Or races, as he called them.

The nineteenth century saw the broad synonymization of these different concepts of analyzing the components of the human species. A race might be a continental subspecies, or a geographical subdivision of one. A race might refer to a linguistic group (like Bantus or Indo-Europeans). Or it might refer to a bounded religious community (like the Amish or Jews); or to a marked ethnic distinction (like the Sami/"Lapps" or Khoekhoe/"Hottentots"). It might refer to a nation (like the French or Japanese); or to physical features (like short-statured Pygmies or depigmented Nordics).

The twentieth century, consequently, saw the gradual recognition that humans assort themselves in all sorts of biocultural ways. None of them is particularly natural or fundamental, and any of them may be very important politically at any given time. Moreover, any individual person has a complex ancestry and multiple concurrent social identities. The assumption that humans come in a few fundamentally different flavors or types, much less that those types could incorporate non-biological criteria, was itself called into question in the latter half of the twentieth century.

The upshot is that race does not have a formal, scientific meaning or application in the study of the human species. And as for its informal meaning, nothing like a few relatively homogeneous natural divisions of our species present themselves. Moreover, as our knowledge of human social processes has grown, we see the reality of race very differently today in science than we did as recently as the 1970s.

Once Again, What Is Race?

All attempts to identify the biological reality of race in the twentieth century failed. A racial scholar of 1900 believed that race was biologically real, as a fundamental organizing principle of the human species, transmitted undiluted from parent to child as a deep internal quality. But as a deep internal essence, race couldn't stand up to probabilistic Mendelian genetics; so from about 1935 to 1960, racial scholars, still believing that race was biologically real, redefined it to be a gene pool. Now you were a part of a race; a race wasn't part of you. But as a gene pool, race couldn't stand up to the empirical structure of human populations, which found them to be not discrete, but overwhelmingly overlapping. By the twenty-first century, a few racial theorists committed to its biological reality tried to redefine race yet again, this time as revealed by the arcane statistical analysis of DNA. But if that's what it takes to determine someone's race, then "race" clearly means something very different and far more subtle now than it used to (see *Understanding Race* by Rob DeSalle and Ian Tattersall in this series).

And yet, race is very much a part of our lives. Particularly if we are non-white. Or more broadly, if we are not a White cis hetero Christian male, because race is actually only one of the many significant identities by which we categorize people, and sometimes do not treat them fairly or open-mindedly on that basis.

Sorting things into categories is a shorthand method for knowing a little bit about them, all other things being equal, and the human mind seems to be particularly adept at it. Race is a sorting device for people – like many others, from nationality to shoe size to musical taste – but this particular one purports to be something scientific. And that represents the big lie of race: That a named group of people can stand as a natural unit of the human species. But there are no such units. As far as we can tell, the closest things to natural units of our species might be living humans, extinct Neanderthals, and possibly also-extinct "Denisovans." But within the extant human species, the subunits are cultural, although they sometimes correlate with biological features.

And why do we believe in race if it is empirically false? Because we believe in many things that are empirically false. Indeed, creating unreal cultural worlds has been the principal way that our ancestors managed to survive in the real ecological world. People die from witchcraft and voodoo, if they believe in them. People die for nationalism, if they believe in it. People have died for love and for God. One of the primary aspects of human culture is that it makes things that are not part of the natural world real; humans thus inhabit a reality made by their ancestors and imposed upon the natural environment that is inhabited by the rest of living things. Race is a part of that cultural reality.

5 Political and Economic Inequality Is Not the Result of Genetics

Race is only significant to human life insofar as it correlates with class, or more broadly, with the quality of life. That is to say, whatever you may consider race to be, it is only a problem if your racial assignment has an effect on your life course. Or, put another way, race exists for racism – more generally, for the reinforcement of political-economic boundaries in society and the maintenance of inequality.

That is a major reason why race can be so confusing. It is social difference masquerading as biological difference. The relationships between colonized and colonizers are frequently reducible to non-Europeans and Europeans. Despite its etymological origin (in the subjugation of the Slavic peoples of central Europe) the relationships between slaves and slavers became easily reducible to Africans and Europeans. And from the other side, as it were, whether they are "gringos," "goyim," "white people," "umlungu," "gaijin," "gwailou," or "sassenachs," these are not simply labels, but boundaries in a social hierarchy in which the speakers may be sneering, but they are also often just a bit envious of the privileged status they are naming. Where the labels denote boundaries of status, power, and privilege, they become effectively "racial" labels without even being officially racial.

The Origins of Inequality

In 1753, the Academy of Dijon posed a question publicly for the scholars of the day: What is the origin of inequality among people, and is it a product of Natural Law? The winner of the contest was Jean-Jacques Rousseau, who argued persuasively that although some people are bigger or stronger or smarter than others, the basic differences between the have-nots and the haves are the products of history, not nature. From an original state of equality, the inequality evident in modern times was due to wheat and iron – that is to say, food production and metallurgy – which permitted some to subjugate others.

This basic idea – that once upon a time people were more or less equal, and that history, not nature, is the reason that they aren't equal any more – became a pillar of the political upheavals that would begin later in the century. Thomas Paine's 1776 screed against monarchy, *Common Sense*, notes that "how a race of men came into the world so exalted above the rest, and distinguished like some new species, is worth enquiring into." And there is certainly no doubt as to where the fault lies: "Mankind being originally equals in the order of creation, the equality could only be destroyed by some subsequent circumstance."

And over the next few decades, that became the starting point for nascent social and political science. What historical forces have brought us here? How can they be undone or modified? Even in his famously turgid 1867 critique of political economy, Karl Marx casually explained, "Nature does not produce on the one hand owners of money or commodities, and on the other hand people possessing nothing but their own labor power. This relation has no natural basis; . . . it is clearly the result of the past historical development."

But there would be a backlash against such analyses that challenged the inevitability of the familiar aristocratic European hierarchy. The person generally credited with first articulating the counterargument – that the rich simply deserve what they have, so go home and STFU – was Arthur de Gobineau, in his 1853 *On the Inequality of the Human Races*. Gobineau was interested in civilization, of which the highest form was Germanic and Christian, and which was thus also conveniently marked out racially. Of 10 civilizations he could identify globally, "all civilizations derive from the white race, [and] none can exist without its help." In fact, imagined Gobineau, civilization waxes with the purity of Aryan blood, and wanes when it is diluted by admixture with anybody else.

Which all implied a moral lesson for the upwardly mobile non-Aryan European of the mid-1800s: Don't even try it. You need the Aryan aristocracy, for they are the bringers and maintainers of civilization. They may often seem like just a bunch of decadent inbred twits, but don't let that fool you: civilization is impossible without them, for it runs in their veins.

The big lesson, though, was about inequality and nature. *Inequality is a fact of nature.* The precise details of that nature may vary substantially, as we'll see, but according to Gobineau the reason that there are rich and poor is not history, not evil doings, not exploitation, not genocide, but biology. The aristocracy is the cream that has risen to the top, by an inexorable unfolding of the innate properties inherent within our species. And by the latter part of the 1800s, the political reactionaries even had a scientific name for the process that had made them rich and most of the rest of the world poor. It was "natural selection," or, if you prefer, "survival of the fittest." "Natural selection" was the phrase for the animal kingdom popularized by Darwin in 1859. A few years later, Herbert Spencer coined "survival of the fittest," and by 1868 even Darwin was synonymizing them. But "survival of the fittest" had a much broader implication than mere "natural selection." Most significantly, survival of the fittest wasn't limited by nature.

Indeed, one could talk about the process by which the rich came to dominate the poor as an extension of Darwinian natural selection. And quite soon, widely read and consequently very influential scientists – like Herbert Spencer in England, Georges Vacher de Lapouge in France, and Ernst Haeckel in Germany – were rationalizing political/economic hierarchies as somehow constituting the inevitable results of the advancement of civilization. And naturally, they selected Darwin's name in which to do it.

The most prominent Social Darwinist in late nineteenth-century America was the Yale political scientist William Graham Sumner. Sumner was not a radical racist or genetic determinist; rather, he was a radical individualist. Sure, it's not a perfect world, argued Sumner in 1881, but the playing field is even, and it's every man for himself.

> The law of the survival of the fittest was not made by man and cannot be abrogated by man. We can only, by interfering with it, produce the survival of the unfittest. If a man comes forward with any grievance against the order of society so far as this is shaped by human agency, he must have patient hearing and full redress; but if he addresses a demand to society for relief from the hardships of life, he asks simply that

somebody else should get his living for him. In that case he ought to be left to find out his error from hard experience.

The attraction of Social Darwinism lay in its convenient sidestepping of the moral question. What moral question? The one considered by the eponymous master himself, as he was rewriting *The Voyage of the Beagle* for its 1845 second edition. While Britain was busily outlawing slavery and America was at least debating it, Darwin was very much the abolitionist: "if the misery of our poor be caused not by the laws of nature, but by our institutions, great is our sin," he mused, and he even went on to argue that slavery is far worse than mere poverty and hardly even comparable to it. Darwin clearly recognized responsibility for the inequalities of the present in the practices of the past, and certainly (as far as slavery goes) he noted that at least We Brits are trying "to expiate our sin."

Of course, confessing and atoning for sins are quite different acts, but Darwin also inadvertently raised a second moral question, which he didn't address. Granting that our ancestors sinned in establishing "institutions" to maintain large disparities in wealth and power, suppose the opposite were true? Suppose that the misery of the poor was indeed caused by *nature*, not by our institutions; then what? Screw the poor, because at least we haven't sinned?

What we have just identified is known generally as the naturalistic fallacy. How we treat the poor should not at all be contingent on how the poor got that way. Misery should be seen as a bad thing, no matter the cause; and citizens and governments should work to alleviate it, no matter why it's there. But this is a moral proposition, not derived from the natural order, and it merits some reflection. Why should it matter if the poor are poor for natural or for unnatural reasons? If the poor and miserable are that way for reasons that aren't institutional, but are in some way biological, does that mean we should feel less sympathy for them? Does it mean that they deserve to suffer?

Yes indeed, says Social Darwinism. And this ignominious strain of political philosophy begins, independently of Darwin and to a large extent contradicting his own beliefs. But its central proposition is this: *Political, social, and economic inequality are due to inherent factors, and are consequently nobody's fault.* To the extent that there is economic stratification in society, it simply is the outward expression of an underlying biological stratification.

The rich have more, and deserve more. Because they're better. And our job as good (right-wing) scientists is to identify just what those biological factors are, which make the rich better and richer than the poor. After all, nobody disagrees that there is inequality; that much is empirical. The issue is the moral one: Given that inequality exists, is it unjust? If the poor are somehow just lazier or dumber than rich, then you can't really blame anybody for their social position.

And thus, when we identify those innate factors, then we can finally blame the inequalities on those factors, and demonstrate that inequality is not injustice, because the rich are entitled to more, by virtue of possessing those qualities that we have identified! On top of which, the rich will also be sinless, as even Darwin appreciated.

So let's get busy!

Heads, Minds, and Genes

The possibility, or expectation, or hope, that there is some sort of naturalistic explanation for the existence of great disparities of wealth and power in European society – and that it isn't due to the inhuman avarice and brutality of the ancestors – emerged as a powerful, if sometimes unconscious, scientific motivation. Anatomists of the eighteenth century knew that it was better to be civilized than to be uncivilized, that civilization comprised a set of ideas, that ideas were located in the brain, and that the brain was located in the head.

They also knew that, architecturally, form follows function. And since the function of a civilized brain is superior to that of an uncivilized brain, it followed that such a difference should be visible in the form of the heads of civilized people. In the late 1700s, the early European racial classifiers certainly believed in the overall inferiority of the non-European peoples of the world, and some even maintained that Europeans had the least ape-like faces (ignoring, obviously, the thin lips and straight hair of the apes). But they didn't yet believe that Europeans had bigger and better heads than the rest of the peoples of the world.

The idea that racial hierarchies could be identified in head sizes was asserted without data by some early nineteenth-century anatomists, empirically refuted by a German anatomist named Frederick Tiedemann in 1836, then re-popularized by the Philadelphia anatomist Samuel George Morton a few years later. After all, maybe a brain is like a muscle, with bigger ones producing stronger thoughts. The idea that political and economic domination might be identifiable in head size indeed became a well-known factoid of early physical anthropology, but over the course of the twentieth century, physical anthropologists showed that: (1) there are large-, medium-, and small-headed people everywhere; (2) head size has a negligible effect on intelligence, except in pathological cases; (3) head size has had no causal relationship to cultural evolution at all for at least the last 30,000 years; and (4) the best predictor of head size in living people is not intelligence, but body size. At least within the range of normalcy, a smart person might have a big head, but a big person always has a big head. And yet, no football interior lineman has ever won a Nobel Prize, despite having such a big brain!

The crude materialism of the idea that the stratification of society might be explained by stratifying the heads of the people in the societies nevertheless remained tempting. By the late nineteenth century the heads of the world were being classified by shape as well as by size (Figure 5.1), and were being

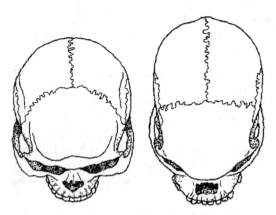

Figure 5.1   Normal human skulls, rounder (brachycephalic) and longer (dolichocephalic).

implicitly invoked to explain opulence at one end and enslavement at the other. "The skull chiefly furnishes the characters of classification; it shows the external shape of the brain, the most important and the highest organ of man; the skull is the means of the classification of the brain," gushed the Italian physical anthropologist Giuseppe Sergi, in 1893.

But by the middle of the twentieth century, head shape was acknowledged to be strongly affected by the conditions of life, sometimes in subtle ways, and sometimes in not-very-subtle ways. Even today, the well-known cranial modifications of ancient South Americans can suggest space aliens to voyeurs who prefer science fiction to science, an idea popularized in 2008's *Indiana Jones and the Kingdom of the Crystal Skull*. The study of skull form has value – not least of all in permitting us to make these generalizations – but its value today is primarily descriptive. As a racial feature, much less as an explanation for economic status, head size and shape simply don't work. The human brain is not like a muscle, in which a bigger and stronger brain generates bigger and stronger thoughts.

If the brain isn't like a muscle, then what is it like? Maybe it's more like an amoeba, whose size and shape can change with the circumstances, while remaining functionally pretty much the same. Or perhaps crude physiology is the wrong place to look, reasoned the American geneticist Charles Davenport, and so he suggested crude genetics instead. Just as Mendel had discovered that elemental traits in peas come in binary forms – wrinkled/round seeds, tall/short plants, green/yellow pods, and the like – we could see people coming in two forms as well, normal and feebleminded. Or, more transparently, smart and stupid. Why are poor people poor? Because they have the gene for feeblemindedness. This idea in fact became quite popular among geneticists in the 1920s, until the Great Depression in the 1930s loosened up whatever relationship they imagined to exist between poverty and genes. Variations on this theme cropped up sporadically in the genetics literature as late as 2005, but without gaining traction.

The most potent scientific weapon against the poor, however, came not from bad craniology or bad microbiology, but rather from bad psychology (see *Understanding Intelligence* by Ken Richardson in this series). Standardized testing, popularized by psychologists during World War I, appeared to demonstrate that the average mental age of American adults is 14. That pseudo-statistic is meaningless, like saying that the average dollar is only worth 78 cents, but it certainly sounded foreboding. But more importantly, the tests clearly showed what Carl Brigham referred to as "the intellectual superiority of our Nordic group over the Alpine, Mediterranean, and negro groups." Never mind that a more careful analysis of those data found that northern Blacks scored higher than southern Whites, and the primary determinant of differences was actually not race, but state expenditure on education.

Eventually the tests were tweaked and rebranded as predictors of scholastic outcomes, rather than as measurements of a subject's genetically encoded mental horsepower. Nevertheless, a small group of right-wing think-tanks and philanthropies continued to promote the lie that brainpower is innate, racially distinct, and readily measurable. The most infamous of these was begun in 1937 and called the Pioneer Fund, which kept a steady stream of money flowing to eugenicists, segregationists, hereditarians, and flat-out white supremacists through the end of the twentieth century. Thus, the scientists on the radical fringe were able to maintain their own journal and to hold meetings from far outside the intellectual mainstream. They managed to keep contact with the mainstream, however, by ostensibly focusing on a fundamentally reasonable-sounding question: Is intelligence at least partly genetic?

Working very hard to show that at least some component of intellectual disparities is innate, the inherent mental superiority of the whitest and wealthiest was championed in twentieth-century Britain by the prominent right-wing psychologists Cyril Burt and Hans Eysenck. Alas, their work is no longer considered even to be honest, much less reliable. Burt actually invented colleagues who publicly aided and praised his own work under their pseudonyms; and dozens of Eysenck's papers have been retracted since his death in 1997. In America, racial ranking by IQ was proposed periodically over the course of the twentieth century, only to be consistently exposed as incompetent scientific reasoning. The most recent example was the notorious 1994 best-seller, *The Bell Curve*, co-authored by a hereditarian psychologist (Richard Herrnstein) and a conservative political activist (Charles Murray).

The Bell Curve

The Bell Curve began with a well-known fact of standardized testing: If you take a thousand Black Americans and White Americans completely at random and give them IQ tests, the distribution of the two sets of scores will be offset by close to 15 points. This is generally understood to be a result of the same phenomenon by which Koreans test an average of 15 points lower than Japanese in Japan. Since Koreans and Japanese test the same in America, the difference is ascribed to the social prejudice experienced by Koreans in Japan.

Without that insight, however, *The Bell Curve* sought once again to explore the origin of the mysterious 15-point Black/White differential. Acknowledging some effects of non-genetic causes upon IQ scores, they statistically found that by very crudely controlling for socioeconomic status, they quickly knocked about six points off of the 15-point difference.

Whereas a reasonable scientific inference might be that more careful controls would get rid of the other nine points, that's not where *The Bell Curve* went. Instead, they claimed to have erased whatever effect the environment might have, and what remained, the nine-point differential, was due to genetics. And so the innate intellectual deficit may be smaller than it used to be, but it is still there, creating a familiar biological barrier to upward mobility.

This was, however, a genetic argument being put forward by non-geneticists, in the absence of any genetic data. The closest *The Bell Curve* ventured to genetics was its argument that intelligence has a significant degree of "heritability." But they used this statistic that they had borrowed from genetics differently than geneticists do. Heritability sounds like it might be a measurement of how "hard-wired," and therefore "inherited," a trait is. But actually that's not at all what heritability measures.

To the geneticist, heritability is a weird and badly named, but importantly population-specific, statistic. Heritability refers to the proportion of phenotypic difference in a population that is associated with genetic difference. It measures correlations, not causes. You can measure the heritability of any trait in any population, but a measurement of heritability in one population is not a predictor of heritability in another population. A strongly hard-wired trait, like the number of your arms, may have a low heritability, since most people with fewer than two have that phenotype for non-genetic reasons. And a non-genetic trait, like whether you are wearing a skirt, might have a high heritability. In fact, that heritability might be significantly different between Americans and Polynesians for entirely non-genetic reasons. If you are wearing a skirt in America, you probably do not have a Y chromosome; but it may not be so clear-cut in Polynesia or Scotland; consequently the measured heritability will differ. *The Bell Curve* falsely argued that since intelligence has a high calculated heritability, any difference in intelligence between two groups of people must be partly due to genetic differences. This argument was in fact just as specious in *The Bell Curve* in 1994 as it was when presented in 1969 by the racist psychologist Arthur Jensen. Geneticists at the time patiently explained it to the right-wing psychologists, but to little avail.

The poor logic that guided the early anatomists' claim to be able to identify civilization, progress, political dominance, and financial status in the size of the head easily becomes recast to focus on the size of the IQ score. We can look at the specious arguments step-by-step.

1. *Economic success is partially mediated by intelligence.* Certainly, there is a broad correlation between money and smarts, but with plenty of exceptions. On one end, the adjunct instructors at universities do a lot of high-end teaching for little pay; and on the other end, fabulously wealthy dunderheads also abound. Obviously, intelligence is no guarantee of anything in modern life.

2. *Intelligence is measured by IQ.* The ability to score well on a pencil-and-paper test is certainly nothing to sneeze at, although it is unlikely to be of much assistance in hunting an antelope, or placating the gods, or weaving a rug, or stealing third base, or changing a tire, or drawing a picture, or frankly, in most human activities. Absolutely nothing in our biological evolutionary history was at all like sitting still for hours on end, trying to focus all of your attention on questions that may seem absurd, filling in the answers with a number two pencil, and wondering what aspects of your life may depend on the answers you give. The greatest ethnocentric conceit of psychometric testing is that intelligence is an individual feature, when the most obvious and time-tested way of finding a solution to a difficult problem is simply to consult someone older and wiser than you.

3. *IQ score is affected by genetic variation.* Whatever those pencil-and-paper tests measure is necessarily, like all phenotypes, the product of genetics and environment. Obviously, people do not possess identical intellectual gifts, and the conditions of life are only a part of the story. But it is a huge mistake to imagine the influence of genes and environment on the human condition to be simple and additive, as if all you had to do was subtract the environment and what's left is genetic. Moreover, the "unit-character problem" was identified by geneticists in the 1920s as the fallacy that genes map on to nouns. The facts that (1) the growth and development of your foot is governed by your genes, and that (2) you have toes, nevertheless does *not* mean that (3) you have "toe genes." Nor do you have chin genes, knee genes, or intelligence genes, despite having a chin, knees, and intelligence. Genes do not have single functions, and what they actually do may be biochemically and physiologically far removed from whatever effect you may be interested in.

4. *Genetic variation explains the IQ differences between populations.* This is the classic sleight-of-hand of confusing the causes of within-group variation and between-group variation. A Navajo and a Rwandan Hutu differ from one another both behaviorally and genetically. Might those genetic differences be causing the behavioral differences? Just a little, teensy bit? Unlikely, because the behavioral and genetic differences are patterned quite differently from each other in our species. People differentiate themselves from their neighbors culturally; cultural variation is predominantly between-group variation. Genetic variation, on the other hand, is predominantly within-group variation, as we have seen. To the extent that there may be genetic variation underlying some behavioral variation, it is likely to be within-group variation as well. In other words, there are depressive, addictive, thrill-seeking, spiritual, smart, dumb, and in-between people everywhere. The question, rather, is: How is such genetic variation distributed? Is it primarily within-group variation, like the rest of the human gene pool, or is this imaginary genetic diversity also supposed to be uniquely and dubiously situated at the margins of cultural groups? Or to think about it another way, imagine an allele with a ridiculously large effect on thought and behavior, say, a "happy" gene that makes you happier than all the people around you. Now imagine that the allele is present in the genomes of that Navajo and that Hutu from the

beginning of this paragraph. Aside from being a little happier than the people around them, sharing that allele would have no convergent effect on any other aspect of their lives. Why? Because living life as a Navajo versus as a Hutu has got nothing to do with genetics, regardless of whether or not we imagine major behavioral genetic variants existing. Similarly, there may well be genetic variation for intellectual abilities, however nebulously defined; but there is also no reason to think that such variation would even be noticeable beneath the huge effects of the conditions of life upon their expression. Thus, different group averages in IQ are incredibly unlikely to be due to the genetic differences that are detectable between the populations.

5. *IQ differences, and ultimately economic differences, between populations cannot be ameliorated, because they are innate.* Like a pigeon heading for home, this is what all the arguments for genetic influences on intelligence or personality ultimately fly toward. The differences are genetic, therefore they are irremediable and immutable, say the radical psychologists. And yet, one of the oldest and best-known fallacies in human genetics is the facile equivalence of innateness and immutability. After all, the very foundation of our evolutionary success has been our ability to transcend what is innate in us, from burning things without possessing heat-vision to flying without wings. What is innate is indeed often mutable; that is the elementary distinction between genotype (what the genetic instructions say) and phenotype (the ultimately relevant parts of a functioning organism). The Social Darwinists were always concerned with traits of imaginary innateness, like laziness and stupidity in the poor. But what about real innate traits, like shortness of stature, or nearsightedness, or cystic fibrosis? Or any other non-innate disability, like a broken leg or toothlessness? The moral point is to make life easier for people with them, not harder. As Thomas Huxley noted, our ancestors were jungle apes, but we aren't. And frankly, what we are learning about apes suggests that they often aren't quite so inhuman either.

6. *Social programs cannot work because they are doomed from the outset.* This is the homing beacon that generated all of the false science. It's what the think-tanks are ultimately paying for: an apparently scientific justification for the social hierarchy we experience. Whether explicitly

racialized or nationalized or not, the goal here is, and has always been, political: upward mobility is bad, stable hierarchy is good, just as it was for Arthur de Gobineau.

Morality and Accountability

The first generation of Social Darwinism barely made it into the twentieth century: a century before *The Bell Curve*, its arguments were widely appreciated to be little more than a pseudoscientific rationalization for the exploitative economic practices of the "robber barons" of the age. That movement was largely replaced in the US by the eugenics movement of the 1910s to 1920s (Chapter 9), with some continuity and differences. The differences between Social Darwinism and the eugenics movement in the US involved (1) the emergence of genetics as the primary science of interest; (2) the recognition that poor people were outbreeding rich people, and its implications for the Darwinian future; and (3) the desire for active government intervention against the poor, in contrast to the Social Darwinists' desire for the government to butt out. The similarity was in the assumption that poor people simply deserve to have little, it's nobody's fault but their own, and consequently that establishing such a pseudoscientific point is somehow more important than alleviating their misery.

After all, wouldn't it be nice to know why they are poor? Isn't that a legitimate scientific question? Why shouldn't we devote some resources and funding towards it?

Let's try and answer those questions with a hypothetical scenario. Suppose that you were granted an audience before the Court of Limited Resources and were challenged with presenting the case for funding the study of the root causes of economic inequality. Why should we devote resources to studying the causes of inequality, asks the Court. Don't we already know what they are? Aren't they – broadly speaking – colonialism, industrialism, slavery, exploitation, belligerence, greed, genocide, and other forms of inhumanity and injustice? And if we do know that these historical processes were the causes, then we should presumably be devoting our resources toward an appropriate response: To try and work for justice and ameliorate the injustice.

On the other hand, you argue in reply, suppose we don't really know what the root causes of social inequality are, and we still think that there is a wee chance that it might be due to internal, rather than external, factors. Perhaps poverty is like a congenital birth defect. Isn't that important to know? Nope, not really, responds the Court. After all, what if poverty were really and truly caused by the feeblemindedness gene? Would that then mean that no amelioration is called for? Would that mean we should be insensitive to people with that condition? Presumably, any sensitive, humane, compassionate impulse would be to try and make life easier, not harder, for them. Sadly, though, the deduction from "internalism" consistently goes in the opposite direction, concluding with the Social Darwinist's "therefore they deserve to suffer."

But they don't deserve to suffer. Nobody deserves to suffer. It's got nothing to do with science; it's got to do with morality. So why does this science consistently lead us to an embarrassingly selfish and cruel place, unless the science itself, in this case, is somehow fundamentally corrupt? If the only people interested in the etiology of class differences in IQ are Social Darwinists, and the only people interested in the etiology of racial differences in IQ are racists, then this is clearly not like other branches of science.

This is a question about the very foundation of science, yet not itself a scientific question. What should we do about evil science? Should research that is intended to worsen people's lives even be permitted?

After all, shouldn't we pause to ask where the roomsful of scientists working for SPECTRE or any other mythical demonic organization actually come from? Is it reasonable to imagine that scientists are so unmoored from morality that they would actually work by the dozens to help an evil organization achieve world domination? Because if you can imagine that, which usually passes unquestioned, then you can begin to grasp the moral problem with the study of the inheritance of intelligence. The study of an organic basis of intelligence is not in itself threatening. But it certainly does not explain economic stratification, poverty, or illiteracy rates better than the history of slavery and colonialism does. Asking whether economic stratification is caused by genetics is like asking whether Jupiter orbits the sun or the earth; we already know the answer. And anyone who thinks otherwise is obliged to confront and acknowledge the political nature of the science they are engaged in, and must be prepared to defend it on that basis. This intellectual turf was never biological; it was always biopolitical.

There is indeed a law of the jungle, but we don't live in the jungle now, do we? – recognized Thomas Huxley, a single generation after *The Origin of Species*. Our job as civilized, post-jungle humans is now "the great work of helping one another." And to that end, argued Huxley, "the ethical progress of society depends, not on imitating the cosmic process, still less in running away from it, but in combating it."

It doesn't matter how natural intellectual aptitudes – whatever those might even be – are distributed in our species. Arguing about them is simply a distraction from Huxley's "great work of helping one another." The fundamental question in bioethics is *cui bono?* – Who benefits? – in this case, by contriving a scientific-looking argument intended to distract you from the "great work." Helping the poor, and working toward equality more generally, should be the end-point *no matter what the cause of inequality may be*. Bringing science into it only makes the science look bad. Especially if the science being brought to bear on the question turns out once again to be racist shit, as it always has.

6 Human Kinship Transcends Genetics

Most primates have a sense of who their mother is. Some have a sense of fatherhood, or at least the-big-guy-who-hangs-around-and-doesn't-kill-me-hood. Many have a sense of siblinghood, or at least of their half-siblings who haven't moved away. Some have a sense of "mom's sister" or other maternal relations, but generally speaking, primate kinship beyond these is subtle at best.

While burning and cutting were certainly crucial technological developments in the ultimate evolutionary success of our lineage, an equally important invention was the family. Unfortunately, social relations are much harder to find in the archaeological record than tools and hearths, so it scarcely gets the attention it merits. Nevertheless, a million and a half years ago, an unusual creature was inhabiting Africa and western Asia. Unlike other apes, it lived on the ground, as its ancestors had been doing with increasing anatomical specialization for a few million years. Also unlike other apes, its canine teeth were small and similar in size between males and females. They were consequently less useful as threats or as weapons, but they did permit a different range of motion by the jaws, and a wider range of sounds for the jaws to produce. The sounds were also related to another distinction from the other apes, an unusually large head.

Of course, we don't know precisely why any of these physical traits initially developed, but by 1.5 million years ago they were very likely working in synergy. The unusual apes had developed a new way of communicating, using meaningful sounds, combinations of sounds, facial expressions, and movements. It was a system that was very effective in permitting cooperation against the hostile forces of nature, but also had to be very slowly learned over a long childhood. This biological strategy of adapting by learning also was accompanied by something that other long-lived, intelligent mammals like elephants and whales lacked: namely, hands. Off of the ground and not being used for locomotion, hands turned out to be especially valuable in transforming local environments. Interacting with their environments, these unusual apes were now approaching the world as if it were composed of things to be altered and used, rather than to just be eaten or avoided. And they used things so well that they fairly quickly spread over an enormous territory, for a species of ape.

This successful investment in learning as a survival strategy came at a cost, though. For one thing, the canine teeth were now involved in vocal communication more than in sexual competition and choice, or for defense. The tongue was also being genetically tweaked for intelligible speech at the expense of its usual ape function, heat dissipation by panting. Moreover, a big-brained baby is harder to expel through the birth canal than a regular ape baby. For yet another problem, a slowly learning child requires a lot more care, and for a longer period of time, than a regular ape baby does. These problems will be solved physiologically and to a great extent socially. Early humans will find distinctly unique ape ways to partner up, to dissipate heat, to facilitate being born, and of course to make sounds.

Unusual Ape Parenthood

Some of the eventual evolutionary tweaks will be biological. The tongue gets a neuromuscular makeover; the skin increases its density of sweat glands and thins the hair follicles, which facilitates evaporative cooling. The baby rotates in the birth canal, and its poorly developed skull bones will bend and slide to facilitate delivery. With the evolution of meaningful utterances comes the co-evolution of meaningful facial expressions. The facial skin will become largely denuded (except secondarily, in adult men) and the eye whites and nasal cartilages will assume greater prominence. The larynx will descend in the throat, facilitating both speaking and choking.

But there is no obvious physiological response to the situation of an incapacitated mother with an underdeveloped hominin (i.e., ancient zoological ancestor or relative) newborn. Where most baby mammals are walking and communicating within hours, not so with these hominins: it will take that baby several years. Moreover, in addition to the underdeveloped neonate, there is a helpless four-year-old to be weaned, and a helpless eight-year-old still hanging on to the metaphorical apron strings. Unlike the human, an eight-year-old chimpanzee is sexually and socially mature. A 12-year-old chimpanzee has its wisdom teeth, and is thus roughly equivalent to a 20-year-old human in dental age; but the 12-year-old human is still far from dental maturity, much less social maturity.

So if hominin mother continues to have babies on a regular ape schedule, about every four years, then she will have a lot more difficulty in handling all those immature children than an ordinary ape mother would. One source of assistance for her will come from a physiological tweak, the development of menopause. An ordinary ape female breeds until she dies; but a human female may still have decades of life after her fertile period ends. What can she do with her spare time? Tend to the grandkids.

Another source of assistance might be from elder siblings. But the one that seems to have been most successful was a cultural invention, like the handaxe, but invisible to paleolithic archaeology: namely, marriage. That is to say, a formalized union between people, which links their families together and creates a social niche for children. Marriage of course takes many forms in human societies, for the simple reason that it has come to serve many functions: establishing a household, legitimizing sexual relations, legalizing children, creating in-laws, and publicly expressing love, among the most familiar. Unlike chimpanzees, who ordinarily mate quite promiscuously, some apes (gibbons and gorillas) form heterosexual bonds, with a female generally socializing and mating with a particular male.

Marriage, however, is different from pair-bonding, in that it is an agreement, not a genetic program. Pair-bonding is instinctual; marriage is contractual. Those aren't necessarily mutually exclusive, but they are different. With marriage, the assistance will come from a member of another family. What marriage might have done, in its primordial form, was to give another source of aid to hominin mother, as a spouse, and to create a network of alliances among families thus united as in-laws. The deal, as it were, would be that hominin mother receives the support she needs while she is incapacitated in child-bearing and overburdened in child-rearing compared with other primate females, and hominin father does not have to compete aggressively to mate, as most other primate males have to.

At least one more physiological tweak would be needed, though. Regular new mother apes act as if all others are potential threats to their infant, which they may well be. But a new human mother can't very well act that way, since she needs someone else there to simply help birth the infant. She doesn't have the luxury of being particularly choosy about anybody else touching her baby. Consequently, the overwhelming protective instincts her ancestors had will have to be dampened, so that the obstetrician, nurse, co-parent, doula, midwife, spouse, sibling, or friend can be there to assist.

The evolution of a spouse is paralleled by the evolution of a sibling. Ordinary ape siblings do not go through sexual maturity together; rather, one sex or the other transfers into another social group. Humans, however, grow physically and socially far more slowly than ordinary apes. We can envision marriage to some extent as a parallel to the sex-biased transfer in apes, but with key differences. First, once again marriage is a legal contract, and consequently can be voided or annulled. Second, sex-biased transfer is largely an individual affair, whereas marriage is much more of a family matter. The scenario of lovers eloping to be an isolated couple is rare – marriage is a cause for public celebration, connecting (at a minimum) two pairs of middle-aged adult parents in a way that is very un-apelike, as in-laws. Third, the relationship that mature opposite-sex human siblings have is quite unique among the apes. Unlike apes, the human brother and sister remain in contact throughout adulthood, as part of a network that embeds their two new families. Yet the contact that they remain in is a very unusual one for adult male and adult female apes: it is non-sexual. In other words, the natural biological relationship of ape siblings is transformed into a special social relationship dominated by a single abstract rule: Love each other, but no sex between you.

In contrast to normal ape social life, then, the social life of our ancestors involved the evolution of three new social statuses: (1) transforming mate into spouse; (2) transforming mother's mother into grandmother/mother-in-law; and (3) inventing brother/sister. These formed the foundation of human kinship.

The Evolutionary Implications of Kinship

At the root of human existence, then, is a novel form of ape social existence, a social existence grounded in abstract, yet local, rules. The rules by which an ape mate is transformed into a human spouse involve a set of expectations and obligations, most fundamentally maternal assistance for mother and non-aggressive paternity for father. These are basically prescriptive rules, expectations about what spouses are supposed to do. The rules governing brother and sister, on the other hand, are dominated specifically by what they are not supposed to do. A marriage of brother to sister will make a statement: "Hey world, we aren't ordinary!" We may be gods or simply perverts, but we are different from you regular folks.

This, of course, would represent the origin of the incest taboo, the rules that bar certain classes of people from being sexual partners, in addition to siblings. The taboo may cover other family members, other household members, or other lineage members, but presumably originated with the need to regulate the sexual conduct of brothers and sisters, who would not grow up together in ordinary ape societies.

The lineage itself is an abstract concept that we probably owe to grandmother. After all, she is post-menopausal and helping to birth and raise her grandchildren. A lineage of three generations presents obvious questions to an abstractive, symbolizing mind: What about dead ancestors before grandmother, and unborn descendants after me? A lineage is there before you and after you, and transcends your own existence and death. Heavy stuff, especially for a brain that asks and answers questions, itself very unusual for an ape brain.

Vertical ancestry or lineage creates horizontal relations or kin in the present generation. Your siblings share parents with you. Your cousins share grandparents with you. Your second cousins share great-grandparents with you. And so on. This actually creates a system in which you know who someone is without ever having met them, again highly unusual knowledge for an ape. However, the genetic relations rarely map on to the social relations. Where one cousin may be a preferred marriage partner, another may be covered by the incest taboo. Indeed, the children of those tabooed opposite-sex siblings may end up happily marrying each other, as cross cousins, like Charles and

Emma Darwin (see below). Charles Darwin's mother (Susannah Wedgwood) and Emma Darwin's father (Josiah Wedgwood) were siblings. The rules vary locally, but everyone knows where they stand.

One relative in particular is very widely singled out for special treatment: Mother-in-law. She doesn't even exist in chimpanzee society, but in humans her presence is felt as much in ethnographies as in situation comedies. As the early anthropologist James Frazer wrote in 1900, "The awe and dread with which the untutored savage contemplates his mother-in-law are amongst the most familiar facts of anthropology." And this, at a time when anthropology had hardly any facts at all!

Mother-in-law and grandmother inhabit the same body, the first in relation to daughter's spouse, and the second in relation to daughter's child. The widespread conflict between spouse and mother-in-law may thus reflect a primordial cultural competition for the same person's affections.

Tweaking the Genetic Relationships

Since father/spouse involves so many different kinds of roles, these can frequently be divided up culturally among different people. Mother's primary sexual partner and children's disciplinarian need not be the same person, as early ethnographers discovered. Breadwinner, household decision-maker, mailbox name, and caregiver can also be divided up in creative ways. And that's without even introducing variations away from monogamy.

Consequently, despite broad similarities, even the labels of relatives may not translate well cross-culturally. Your "uncle" may be a blood relative (your father's brother) or a non-blood relative (the man married to your father's sister). It's actually pretty weird to think that these two people should be called the same thing, and certainly goes against the genetics. Conversely, the child of your father's brother and the child of your father's sister may both be cousins to you (and to the geneticist), but may be regarded quite differently in different cultures (as a parallel cousin and a cross cousin, respectively – as noted above).

The biologist J. B. S. Haldane famously remarked, in anticipation of modern kin-selection theory in biology, that he would give his life for two brothers or eight cousins. This reflects the amount of DNA that would be mutually inherited from a common ancestor, and therefore identical by descent. In reality, though, Haldane only had one sister and three cousins. Does that mean he would have let them drown, and not taken a risk to save them?

Of course not.

Well, hopefully not, at least. But it does show that even though Haldane thought as a geneticist in theory, in practice he probably would have thought differently, as a normal compassionate human being. When the chips are down, even geneticists probably don't think like geneticists. Genetic relations are often incorporated into patterns of kinship, but they also contradict patterns of kinship. In the simple case just noted, calling your father's brother and your mother's sister's husband the same thing ("uncle") and treating them the same way actually defies their genetic relations to you.

Americans are generally uncomfortable with the idea of a second cousin as a spouse, but in fact that consanguineous marriage is legal in all 50 states. The creepy feeling of an "incest taboo" covers culturally assigned relations, not genetic ones, and not necessarily even legally acceptable ones. You might consider someone who shares your last name as kin, even though you've never met them before. You might consider someone who has the same rare genetic allele as you to be kin. You might consider an adopted relative to be kin. You might consider an in-law to be kin. You might consider step-relatives to be kin. All of these statuses are uniquely human, and serve to show the broad area in which human kinship and genetic kinship don't necessarily map on to one another.

The most basic realities of human kinship – aside from the trivially obvious proposition that we tend to support our kin over other people – contradict the genetic relations. After all, just three generations ago, you had eight great-grandparents, and you quite possibly can't even name them all. To simplify ancestry, most societies choose one sex (matriliny) or the other (patriliny) to track ancestry.

The ancient Hebrews, for example, were famously patrilineal. A wife joined her husband's family, and the children they had together belonged to his lineage. Consequently, the ancestry of Jesus as given in Luke's gospel takes us back through his presumptive father Joseph, his father Heli, his father Matthat, and 17 more generations back to a Hebrew leader named Zerubbabel, who is mentioned several times in the Old Testament. It consequently appears to the reader as though Jesus had only one lineal ancestor 20 generations earlier, when he actually would have had over a million (assuming complete outbreeding). What this actually tells us is that of the million or so ancestors in the 20th generation, only one actually matters: his father's father's father's . . . father. This is a lineage, literally linearizing the network of ancestors, and reducing it to a manageable unidimensional scale.

And incidentally completely contradicting the biology of ancestors and descendants, as we will see. After 20 generations of Mendelian genetics, Jesus would have had exceedingly little, if any at all, of Zerubbabel's DNA.

Fictive Ancestry

But there is a bigger problem when you confuse what people say about ancestry with what biology says about ancestry. To continue with the example of the canonical biblical ancestry of Jesus, we have been citing Luke's gospel, which gives the relevant part of the ancestry of Jesus as: Zerubbabel ➔ Rhesa ➔ Joanan ➔ Joda ➔ Josech ➔ Semein ➔ Mattathias ➔ Maath ➔ Naggai ➔ Esli ➔ Nahum ➔ Amos ➔ Mattathias ➔ Joseph ➔ Jannai ➔ Melchi ➔ Levi ➔ Matthat ➔ Heli ➔ Joseph ➔ Jesus. These are all Hellenized versions of presumably Semitic names and fathers.

The gospel of Matthew, however, tells the same story rather differently, or perhaps tells a different story altogether. According to this canonical scripture, the relevant part of Jesus's ancestry from Zerubbabel goes like this: Zerubbabel ➔ Abiud ➔ Eliakim ➔ Azor ➔ Zadok ➔ Achim ➔ Eliud ➔ Eleazar ➔ Matthan ➔ Jacob ➔ Joseph ➔ Jesus. Not only don't the names match, but given the generation length of human beings, Luke's genealogy of Jesus would have to entail a few more centuries than Matthew's genealogy of Jesus.

Obviously, to the modern reader the genealogies are the products of two different early Christian communities, both of which felt the need to establish the descent of Jesus as the Messiah from the lineage of the ancient King David. Zerubbabel was a "known" descendant of King David, named in several ancient books.

All they had to do was to come up with a bunch of names linking Jesus's presumptive father, Joseph, to biblical Zerubbabel. And so they did, although with wildly different names and numbers of generations. And it's only a problem if you think about it in a narrow naturalistic way. It is probably best understood as a reasonable response to the theological necessity of linking Jesus to King David, and filling in the parts that they didn't know.

In addition to misleadingly linearizing ancestry, it is a common human practice to simply make up remote ancestries. After all, who is really going to be able to deny your progenitors 12 generations back? In the modern age names are commonly altered to conceal ancestries that are not sufficiently Anglo-Saxon (like Izzy Demsky, who became Kirk Douglas) or sufficiently euphonious (like Frances Gumm, who became Judy Garland). Humble or embarrassing ancestries can be concealed by a name change, as John Rowlands became Henry Morton Stanley ("Doctor Livingstone, I presume?"). Difficult-to-pronounce-or-spell names in indigenous languages may be transformed into more manageable ones in colonial circumstances, leading to confusion generations down the road. I might be related to people named Marks or Markowitz, or Markovich, and we might all be descended from somebody named Mark.

Or perhaps Bob. It doesn't really matter.

Family surnames are a fairly recent imposition upon the peoples of the world, for the mundane bureaucratic purposes of census and taxation. Even so, the common Jewish surname Katz is actually a contraction of the Hebrew for "true Cohen (i.e. priest)" – which in turn may imply an old problem with fake Cohen/priests out there.

Different Ancestries

Genetics is often better at predicting the future, which involves probabilities, than at reconstructing the past, which involves pattern recognition. There are, after all, multiple possible genetic futures, but only a single actual genetic past. To understand the complexities of genetic ancestry, we need to revisit the two basic principles of chromosomal inheritance. First, you have two versions of each chromosome, one from father's sperm and the other from mother's egg,

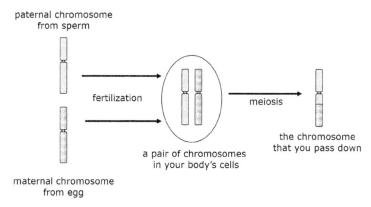

paternal chromosome
from sperm

fertilization

meiosis

a pair of chromosomes
in your body's cells

the chromosome
that you pass down

maternal chromosome
from egg

Figure 6.1 Every generation a particular chromosome decays due to crossing-over during meiosis.

but you pass on only one of them to each of your offspring. And second, the chromosome you actually pass on is neither your mother's nor your father's, but a combination of both.

Let us consider the intergenerational transmission of a single chromosome. You inherited one version of a chromosome in your mother's egg, and another version of the same chromosome in your father's sperm. As illustrated in Figure 6.1, the chromosome that you pass to your child is not identical to either of the ones you inherited, but is rather a mixture of the two, due to crossing-over during meiosis, the cell division going on in your gonads (see below). Crossing-over seems, at least at first approximation, to be able to occur anywhere on a chromosome; and it must happen once, but may happen a few times on any specific chromosome. In other words, crossing-over seems to happen a low, but random, number of times, at mostly random places along each chromosome. Meiosis is a precisely ordered system with a great deal of randomness built in.

The randomness of chromosomal segregation and crossing-over is what makes the genetic future statistically predictable, and the genetic past paradoxical. The basic rules for statistically predicting the genetic future were worked out in fruit flies about a century ago, and apply remarkably well across multicellular life. The paradox of the past has only been explored more recently and is based on a principle known as pedigree collapse.

You had two parents. They each had two parents, your four grandparents. They each had two parents, your eight great-grandparents They each had two parents, your 16 great-great grandparents. They each had two parents, your 32 great-great-great-grandparents. And so on (presumably back to the origin of sexual conjugation a billion or so years ago, a freaky thought, so let's not dwell upon it). There is a simple formula for calculating the number of genealogical ancestors you had in any ancestral generation. If you go n generations back, you had 2^n ancestors in that generation. Thus, one generation ago, you had two parents, and five generations ago, you had 32 great-great-great-grandparents.

Moreover, just like Jesus, as noted earlier, you would have had over a million ancestors just 20 generations ago. However, those million lineal ancestors of 20 generations ago would not have all occupied different bodies. The reason is statistical and random: all human populations are finite, not infinite in size, and as such they undergo some degree of inbreeding. Inbreeding creates a discrepancy between the number of your genealogical ancestors (the ones in your genealogical pedigree) and genetic ancestors (the ones whose DNA is perpetuated in your own cells).

Consider the offspring of a union of first cousins, which is ethnographically widespread, and was particularly well known as way to keep the money "in the family" among the wealthy classes in Victorian England. Charles Darwin and Emma Wedgwood were first cousins, and they had 10 children together. Let's look at the ancestry of one of them, Leonard Darwin (Figure 6.2). Leonard Darwin had two parents (Charles and Emma Darwin), and four grandparents (Robert and Susannah Darwin, and Josiah and Elizabeth Wedgwood). But three generations back, he had six, not eight, great-grandparents. Why? Because his grandmother Susannah and grandfather Josiah were themselves siblings. Two people chosen at random have four parents, but two people who are brother and sister only have two parents. Each of them would essentially represent two genealogical ancestors inhabiting a single body, a genetic ancestor.

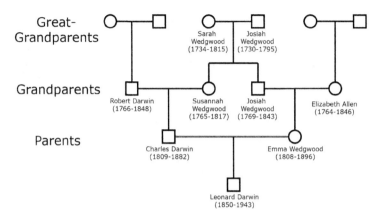

Figure 6.2 Ancestors of Leonard Darwin, whose siblings are not shown. Because two of his grandparents were siblings, he only had six great-grandparents, rather than eight. He also was head of the British Eugenics Association (Chapter 9) from 1911 to 1928.

Francis Galton, another member of this inbred clan, began calculating in 1890 the contribution that each great-grandparent should contribute to a child. Statistically, we would now readily acknowledge that each of the eight great-grandparents should contribute one-eighth of the ancestry of a child. But clearly, Sarah Wedgwood and her husband Josiah are each contributing two-eighths to Leonard Darwin's ancestry, their chromosomes passing through both Susannah and Josiah II on their way into Leonard's cells.

So the number of genealogical relatives overestimates the number of genetic relatives, because some genealogical relatives have to be counted more than once in any inbred pedigree. And now the bad news: All human populations are inbred to some extent, for the simple reason that they are finite in size. And the more finite they are (i.e., the smaller the populations are), the more inbreeding there is going to be. Not necessarily by choice, as in the Darwins and Wedgwoods, but just by the simple mathematics of small gene pools.

Moreover, the farther back in time we go, the more inbred we realize we are. Once again, the product of a numerical paradox. Consider a modern living genealogical descendant of Brian of Nazareth, who may have been a contemporary of Jesus, and may have had offspring. How many generations ago was that, separating the early Roman empire from the late American empire? Assuming four generations per century and 20 centuries, that would come out as 80 generations. Brian would have been just one of 2^{80} or something like a septillion genealogical ancestors in his generation, of that living descendant. Of course, that is many many orders of magnitude larger than the number of people alive at that time. In fact, it's many many orders of magnitude larger than the number of people alive today, which is more than there ever have been at any time in the past, because, as we all know, the human population has been increasing in size through time.

So as you go back in time, the number of your genealogical ancestors is doubling every generation, but the number of people alive at any time is diminishing. Now we must use computer simulations to reconstruct the gene pools of the past rather than the classic probabilistic statements about the gene pools of the future. But there are many possible futures, and only one past – the question is how to find it. As you enter a genealogical time machine and head back into the past, the problem you encounter is having to cram more and more genealogical ancestors into fewer and fewer ancestral bodies.

That is why an amateur genealogical study in 2012 claimed to be able to show that all of the US presidents except one – Martin van Buren – were lineal descendants of a common thirteenth-century ancestor, King John (Lackland) of England. Since each of those 44 presidents had billions of lineal ancestors back in the thirteenth century, it shouldn't be terribly surprising that there would be a bit of overlap among them. That was only 800 years ago. And the farther back you go, the more overlap there must necessarily be among the ancestors of anyone alive in the present. Eventually, as you travel back through the generations you arrive at a point many generations ago, in which everyone alive at that time would be either the ancestor of (1) nobody alive today, since a lineage can always die out, or (2) everybody alive today. In other words, everyone alive today would be drawn from the same pool of genetic ancestors alive back then. The only difference among living people would be how many times each genetic ancestor back then recurs in the genealogical ancestry of each living person today.

And that time turns out to have been surprisingly recent, perhaps the late Pleistocene (which ended 11,700 years ago). In other words, someone alive say, 25,000 years ago would be mathematically constrained to be either everyone's ancestor in the twenty-first century of the Common Era or nobody's ancestor.

That is the ultimate result of looking back through time at the collapse of the supergazillions of genealogical ancestors into the perhaps millions of people alive about a thousand generations ago. Of those millions – spread across Africa, southern Europe, and southern Asia – some are destined be the ancestors of none; some are destined to be the ancestors of all; but none will be the ancestors of only some.

That fact might form the basis of an ad campaign for racial tolerance, but of course we cannot validly derive moral facts (like political equality) from natural facts (like a recent human genealogical coalescence date). In fact, it may still be unclear just what we mean by a recent human genealogical coalescent date (given a measured mutation rate, the projected date in the past when all the present-day genetic variation under consideration is deduced to have originated), much less to derive a moral precept from it.

Crossing-Over

The process of cell division known as meiosis involves halving the number of chromosomes per gonadal cell, so that fertilization can restore the ordinary (diploid) chromosome number. This involves sorting a random member of each pair of chromosomes into a daughter cell, and it forms the basis for Mendel's discoveries of how traits sort in generations of pea plants. However, the early studies of genetics in fruit flies showed that entire chromosomes were not in fact being transmitted intact into the sperm/egg/pollen/ovule. Rather, at the beginning of meiosis, each particular chromosome identifies its mate or homolog (we still don't know how!) and pairs with it, and exchanges bits with it. This is actually visible under the microscope, and has come to be known as crossing-over. The result of crossing-over is that the chromosome #3 in your sperm or egg is not precisely the same chromosome that you inherited from either your father or your mother. Rather, that chromosome #3 is composed of parts of your father's chromosome #3 attached to parts of your mother's chromosome #3. In this way, genetic variants (alleles) physically enter into new combinations every generation.

Crossing-over was discovered several years after the details of chromosomal segregation had been worked out. It was initially regarded as an exception to the rules, or even laws, of chromosomal segregation, but we now know it to be a critical part of the normal operation of meiotic cell division. When crossing-over fails to occur, that's exceptional, and bad.

Since crossing-over seems to occur in random spots along the chromosome, its long-term effect is to chop an ancestor's chromosome into little bits, as it travels through the generations into the bodies of descendants. Let's look at the transmission over the generations of the chromosome we saw earlier. The chromosome that you pass on to your child is actually neither mom's nor dad's, but a bit of both, combined, because of crossing-over in your own gonads. We can't predict specifically which chromosome bits will be transmitted to a particular child, but we know that, on average, 50% of what you transmit chromosomally to your child will be from your mother and 50% of what you transmit chromosomally to your child will be from your father. Of course, because of the randomness involved, in any particular case, it might be 45%. Or 55%.

A generation later, your child will have a 50:50 chance of passing on that particular chromosome. But once again, that chromosome won't be intact, for it will be composed of random chunks of three chromosomes all joined together. The chromosome we are tracking, which you inherited from your parent, continues being whittled down over the generations by crossing-over (Figure 6.3). If you pass on that chromosome to your child, it's going to be, on the average, maybe about 25% identical to the chromosome you inherited, but again, it could be 20% or 30% identical, depending upon the randomness in the number and locations of crossovers. Let's track it for one more generation.

That chromosome again has a 50:50 chance of getting into the next generation, but now it is composed of random chunks of four chromosomes. The chromosome you received from your parent again is being whittled down, with only a 12.5% chance of any particular genetic bit having been passed on intact from your parent to your grandchild. What then, is the overall pattern we find in the transmission of chromosomes as a result of crossing-over?

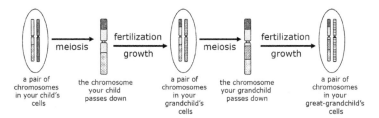

Figure 6.3   Any particular chromosome will be whittled down by crossing-over every generation. And consequently, after a few generations very little of the original chromosome will remain in a descendant.

In any generation, the chromosome being tracked will be composed of a large chunk of a different chromosome (from a new crossover) and an approximately equally large chunk of the whittled-down chromosome from whatever distant ancestor piqued our interest initially. Now, let's say that that distant ancestor lived around 11 generations ago, and so that chromosome chunk will be composed of bits of 10 other chromosomes which had crossed-over during the previous generations. Since you have 2^{11} or 2,048 genealogical ancestors 12 generations ago, we can safely say that any particular bit of chromosome from the distant ancestor has a <1% chance of showing up in the cells of their distant descendant. Or, to put it another way, any lineal ancestor in that generation should have contributed, on average, <1% of any of your chromosomes. But because of all the statistical randomness along the hereditary path, that ancestor might have contributed 2% of the chromosome, or 0%.

And once that contribution drops randomly to zero, it represents yet another genealogical ancestor who isn't a genetic ancestor.

Relatedness Is Constructed

Looking at human kinship in solely biological terms is exceedingly unproductive, because it ignores what makes human kinship unique and interesting. Human relations are biocultural, in that they are partly formed by biological heredity, but also by non-biological rules of exclusion and inclusion, which may frequently override the biological relations.

Your distant ancestors may actually have no particular genetic continuity with your own genome; and there is no guarantee that your recent ancestors passed on their very best bits of DNA to you. That creates immediate ambiguities about the genetic connections between ancestors and descendants. And need we even go into the possibility of false paternity somewhere in the lineage? Even the gospels of Matthew and Luke acknowledge that the (irreconcilable) genealogies they are presenting are the ancestries of Joseph, not of Jesus, and yet Joseph was presumably not the actual father of Jesus. It indeed poses a logical conundrum to present a patrilineal ancestry of someone whose father was a Holy Spirit.

And the most remote ancestors are notoriously easy to tweak creatively. Whether you claim descent from Charlemagne, the Eagle Spirit, or mitochondrial Eve, who can really say that you aren't descended from them? The Gospel of Luke, after all, is not satisfied with relating the patriline of Jesus back to King David, or even back to the hazier patriarch Abraham, but takes him all the way back to Adam. Whatever we can say with any degree of certitude about Jesus, it is probably a bad bet to try and connect him genealogically and/or genetically to Adam and Eve (particularly since Adam and Eve, according to the Bible, would have been genetic clones of one another).

7   Men and Women Are Both from Earth

On April 4, 2022, Congressman Madison Cawthorn of North Carolina, a conservative firebrand, delivered a speech intended to lecture Democrats on the meaning of "woman." A woman, he said, is "XX chromosomes, no tallywhacker. It's so simple."

Chromosomal sex is of course one facet of human sexual biology. Ordinarily, someone with a large X and a small Y will be born male, and someone with two large X chromosomes will be born female. By "born" male or female, we mean that they will be assessed as one or the other, usually on the basis of possession of a tallywhacker (i.e., a penis, to adults), or lack thereof; and will be raised and socialized accordingly. Chromosomal sex, however, is hardly uniform across species. In birds, for example, the system is essentially reversed, so that the female is the one with two different sex chromosomes and the male is the one with two similar sex chromosomes.

Although the XX/XY system is normative in the three best-genetically studied species – fruit flies, mice, and people – it is not universal simply across the mammals, and the superficial similarity often conceals profound differences. Most notably, sex determination in the fruit fly is based on the number of X chromosomes present, while sex determination in the human is based on the presence or absence of a Y chromosome. How do we know this? Because an XXY fly develops as a female, while an XXY human almost always develops as a male. That obviously implies that the overall physiology of sex is probably a bit more complex than it may seem at first blush.

In humans, a gene on the Y chromosome called *SRY* releases a biochemical signal for the developing embryo to begin developing into a male. If, for some mutational reason, the gene is not present or not functioning, then the embryo will develop into an apparent female in spite of having an XY sex chromosome configuration. Likewise, if the signal is released but not recognized by its physiological target, the XY person will develop into an apparent female. In such cases, and others like them, in which the molecular endocrinology doesn't quite work properly, the chromosomal sex will stand in opposition to the sex assigned at birth and lived by the subject.

Other aspects of sex may not match up against the human chromosomes, either. Not only is there the sex you are assigned at birth and are raised as, but there is also the sex of your gonads, the sex of your post-pubertal body, the sex you identify as, not to mention the sexual partners or acts that you may find most stimulating, any of which may not coincide with the rest, with attendant ambiguities. And that isn't even counting the biological intersexes and hermaphrodites.

Had the (now ex-) congressman known at all what he was talking about, he would never had made such an idiotic comment, especially the part about it being simple. To define a woman in terms of simply chromosomes and penis is like defining a bird as a feathered, flying animal. It works to a first approximation, unless you know about the existence of, say, penguins, ostriches, and domesticated turkeys. Presumably knowing about the existence of birds like those would be a motivation for revising your definition of birds, and recognizing it as a bit less simple. As they say, a little knowledge is dangerous; and very little knowledge is very dangerous.

Today we think of sex as not so much binary as bimodal, with plenty of people between, or crossing, the two predominant categories.

Congressman Cawthorn had prefaced his mercifully brief lecture with "You'll never amend biology. Science is not like Burger King. You can't just have it your way." But of course biology is being continually amended; that's why Aristotle is required reading in Classics and History today, and not in Endocrinology. Indeed, an acknowledgment of human evolution on Cawthorn's part would certainly have been a more welcome foray into science.

Charles Darwin

Perhaps the least well-kept secret in science is that Charles Darwin was a wealthy Victorian gentleman, with the social values and prejudices that generally accompanied his era and class. Since he deliberately scrubbed *The Origin of Species* nearly clean of any references at all to people, those values generally aren't visible in that book. But when Darwin published *The Descent of Man* in 1871, they became more evident, if not always to his contemporary readers, then certainly to readers a century and a half later.

To be sure, Darwin was on the socially progressive end of wealthy Victorian prejudices. His German correspondent, admirer, and ally, Ernst Haeckel, had argued in 1868 that there were 12 species of living humans, divided into 36 races, only one of which was actually fully human. As a question of zoological taxonomy, argued Haeckel, "the most highly developed and civilized man" is distinct from "the rudest savages," who "have to be classed with the animals." Darwin himself believed that there was but a single human species – clearly an open question at the time within the scientific community – but he never actually argued the point with Haeckel. Darwin was appalled by slavery, and horrified by the conditions imposed upon indigenous South Americans. He even took taxidermy lessons in Edinburgh from "a negro" named John Edmondstone, who was "a very pleasant and intelligent man."

Nevertheless, Darwin says near the beginning of *The Descent of Man* that Haeckel's recent work:

> fully discusses the genealogy of man. If this work had appeared before my essay had been written, I should probably never have completed it. Almost all the conclusions at which I have arrived I find confirmed by this naturalist, whose knowledge on many points is much fuller than mine.

And by the end of the work, Darwin closes the distance between himself and Haeckel even further, explaining that

> the Negro and European, are so distinct that, if specimens had been brought to a naturalist without any further information, they would undoubtedly have been considered by him as good and true species.

Darwin is judiciously mute about classifying the rudest savages with the beasts. But when it comes to women:

> The chief distinction in the intellectual powers of the two sexes is shewn by man attaining to a higher eminence, in whatever he takes up, than woman can attain – whether requiring deep thought, reason, or imagination, or merely the use of the senses and hands.

And how did this obvious chief distinction arise?

> To avoid enemies, or to attack them with success, to capture wild animals, and to invent and fashion weapons, requires the aid of the higher mental faculties, namely, observation, reason, invention, or imagination. These various faculties will thus have been continually put to the test, and selected during manhood. Thus man has ultimately become superior to woman.

Darwin was not trying to prove that women are intellectually inferior to men; he was simply assuming it, because it was obvious, and trying to come up with a biological explanation for it. Indeed, he even acknowledged that he was not quite such a progressive thinker on the subject of women's intellectual faculties: "I am aware that some writers doubt whether there is any inherent difference, but . . . "

In fact, however, Darwin's sexism left his work vulnerable to attack by the (early feminist) creationist William Jennings Bryan some decades later. "If [Darwin] had lived until now," wrote Bryan in 1922, "he would not have felt it necessary to make so ridiculous an explanation, because woman's mind is not now believed to be inferior to man's."

So a century and a half after publishing *The Descent of Man and Selection in Relation to Sex*, Darwin seems to have been a bit less racist than some, and a bit more sexist than others. It isn't clear what effect, if any, this should have on his legacy as a scholar and scientist, any more than the knowledge that Plato and Aristotle were anti-democracy might affect their legacies. Darwin, however, looms large in the modern scientific pantheon as a cult figure, any criticism of whom strikes a vicarious blow for creationism. Consequently, when *Science* – the leading science journal in the US – published an editorial on 150 years of *The Descent of Man*, anthropologist Agustín Fuentes wondered what the lessons should be when we have students read Darwin:

> Today, students are taught Darwin as the "father of evolutionary theory," a genius scientist. They should also be taught Darwin as an English man with injurious and unfounded prejudices that warped his view of data and experience. Racists, sexists, and white supremacists, some of them academics, use concepts and statements "validated" by their presence in "Descent" as support for erroneous beliefs, and the public accepts much of it uncritically.

That seems, as the editors at *Science* apparently deemed it, to be pretty uncontroversial stuff. Of course, both Darwins inhabited the same body, and obviously failing to examine and criticize the bad ideas that Darwin articulated is pedagogically problematic both scientifically and morally. But the reflex action against the essay was swift: "We fear that [it] will encourage a spectrum of anti-evolution voices." Apparently, to the Darwinian fundamentalists, creationism is worse than scientific sexism and racism.

Permit me to disagree. Let us rather teach students to identify and purge the outmoded ideologies of sexism and racism from science with the same fervor as they identify and purge creationism, and science will be in a much stronger moral position in society.

Gender

Early twentieth-century scholars had begun to call attention to the distinction between femaleness and femininity, or between the natural and cultural aspects of being a man or woman. By the 1970s, this distinction became crystallized as "sex" and "gender." The former would refer to the biological or naturalistic aspects of being a man or woman, while the latter would designate the expectations of society, or the roles one grows into. It represented an expression of the primitive geneticist Francis Galton's segregation of "nature" from "nurture," and the primitive anthropologist Edward Tylor's distinction of "race" from "culture." It was intended to capture the differences among peoples that weren't innate, but were instead historically produced: mode of speech and adornment, posture, diet, attitude, or in Tylor's words, "knowledge, belief, art, morals, law, custom, and any other capabilities and habits acquired by people as a member of society."

The cultural aspects of human population differences came to be crudely labeled as ethnicity. In other words, ethnicity was intended to mark the differences among human groups that are not natural, or racial, or genetic, or biological. We now see the dichotomy as being a bit false, but to a first approximation at least, the distinction was critical by the 1960s. It permitted scholars during the Civil Rights movement to distinguish between the known physical differences between Blacks and Whites, and the differences attributable to the legacy of systems of slavery and segregation, or even just centuries of slanders. This also permitted the Irish and the Jews, traditional racial identities in Europe, to be submerged within the various other immigrant American ethnic communities over the course of the twentieth century.

We now see race and ethnicity as less distinct, and slightly more permeable, but the fundamental distinction between the differences ascribable to a transcendent innate nature (race) versus those ascribable to malleable culture (ethnicity) remains crucial. It remains crucial because it gave legal scholars a conceptual apparatus for dismissing the argument that economic and social inequalities are the consequences of biological differences. Biological differences among groups may be quite real, but nevertheless became scientifically non-explanatory for the origin of political-economic inequality over the course of the twentieth century, either internationally or domestically.

In parallel, consequently, feminist scholars of the 1970s, likewise arguing for equality, began to adopt "gender" as a learned, cultural antonym to innate, biological "sex." And just as racism, not race, was responsible for Black/White inequalities, so too was sexism, not sex, responsible for man/woman inequalities.

By the end of the twentieth century, however, it was clear that the relationships were rather more complicated than could be captured in simple binaries. Race was no longer seen as the biological structure of the human species, and sex itself was being seen as less binary. Both turned out to have aspects that were significantly, if not obviously, cultural. Race could still stand for the biological correlates associated with discrete social and political distinctions; and ethnicities widened to encompass nearly any of the myriad symbolic ways in which humans group themselves: "based on the race, region, religion, sect, language family, language, dialect, caste, clan, tribe or nationality of one's parents or ancestors, or one's own physical features."

Similarly, sex:gender began to break down, as sex became more biologically complicated than the simple binary, and gender expanded to include self-identified LGBQTI (lesbian, gay, bisexual, queer, transgender, intersex) people. Nevertheless, the distinction it marks is still crucial. Some innate biological differences between men and women are certainly real, but they don't explain the political-economic differences that they may correlate with.

Human Sexual Dimorphism

It isn't that difficult to tell a male from a female gorilla. The male is the big one. If you only have their skeletons, he's the one with the bigger bones, and the bigger ridges, anchoring the muscles to move those bigger bones. And if you only have their skulls, he's the one with the gigantic canine teeth, and the muscles and bony crests to move those jaws.

Human males and females are not quite so easy to tell apart. Indeed, the ways that men and women differ from one another often has no homolog in our closest relatives, the apes. Sure, you could look at their relative body sizes, but there is quite a bit more overlap between male and female people than between male and female gorillas. Men are on the average slightly bigger and stronger than women, but whatever your size or sex you'd have to be crazy to pick a fight with Ronda Rousey (the charismatic mixed martial arts Women's Bantamweight champion, 2012–2016), or Valentina Shevchenko (the Women's Flyweight champion, 2017–2023)

Your best cues would be from the features that have no counterpart in the apes: gender. Of course, if you don't know the local rules by which men are supposed to decorate themselves or move about differently from women, then you'll have a harder time of it. So you might rely on a set of physiological criteria, which also have no homologs in the apes: The one with the beard and prominent tally-whacker is probably the male, and the one with the breasts is probably the female. Both are highly variable features, but with rather little overlap.

And suppose you only had their skulls? The one with the bigger face and larger brows is probably the male. Unlike the apes, their canine teeth won't tell you much. But also unlike the apes, there is a bump behind the ear, known as the mastoid process – because it reminded some early anatomist of a breast – and which is rather paradoxically somewhat larger in men than in women, but for no obvious reason. Actually your best bet for sexing a human skeleton would be from the pelvis. The one that looks like a cantaloupe-sized head might be able to pass through it is the female. Not so for the ape's pelvis, however. The female ape's pelvis doesn't face so extreme a challenge, and consequently doesn't have any anatomical specializations to facilitate parturition. The male and female ape pelvis look pretty much the same (except that the male is the bigger one).

There is an obvious lesson in all of this: Whatever processes are responsible for sexual dimorphism in the apes were operating quite differently in our hominin ancestors, and produced quite different results. There is a bit of continuity with the apes in body-size dimorphism and genitalia, and a great deal of discontinuity in the boobs, butt, face, overall body composition, hairiness, pelvis, mastoid process, and maturation process. That of course is worth considering when confronted with the argument that the same processes of sexual selection that operate in apes have also operated in us. After all, the same processes of electromagnetism have operated, and with similar results, in apes and people; but in the case of sexual selection the outcomes of the relevant processes involve the production of differences that are much more striking than the similarities. Consequently, we must imagine different forces at work in producing the patterns of sexual dimorphism in apes and people.

Darwin actually had two theories of competition developed when he wrote *The Origin of Species* in 1859, but he only discussed one: natural selection. This was the competitive process by which an ancestral species would develop into two divergently adapted descendant species. Darwin thus managed to develop a naturalistic explanation for biological adaptation, which had eluded scholars since at least Aristotle. Having explained why closely related species look different from one another, Darwin was left with another problem: why members of the same species, namely males and females, often look different from one another. In both cases, it was competition. In natural selection it was competition to survive; in sexual selection, it was competition to breed. Moreover, since females can generally find a breeding partner without much trouble, it's the males who have to compete for females, thus stimulating evolution in males more than in females – hence the lion's mane and the peacock's tail. Once again, it was something nobody had ever successfully explained naturalistically before.

The trouble was that it projected all of nature through a Victorian lens; and when it came to humans, the best that sexual selection theory could do was to universalize Popeye and Bluto fighting for the affections of Olive Oyl, who will passively accept whoever wins, even if she does like Popeye a little better. But imagining human sexual evolution in such a fashion would result in a "woman that never evolved," as labeled by anthropologist Sarah Blaffer Hrdy in the 1980s – that is to say, a woman like Olive Oyl, seemingly produced by evolution to be passive, submissive, and helpless. That is not at all how women evolved, argued Hrdy. A properly Darwinian, or perhaps neo-Darwinian, or perhaps post-Darwinian, view of the evolution of women acknowledges the ways in which a female human is primatologically not like a female ape as seen through a lens of Victorian gender stereotypes. Rather, she is the result of a broader understanding of the processes of sexual selection.

Even Darwin recognized that females aren't entirely passive, but can be fickle, thus stimulating more intense competition among the males. But what would our female ancestors have been selecting males for? Certainly not their canine teeth. Their whiskers? Why on earth would they be interested in that? And why would a hominin male desire his females to be plumper than an ape male desires his females to be? Why would a hominin female desire a male with a large mastoid process, which isn't even visible externally?

One possibility is that the features we observe don't map on to the features that nature produces – the unit-character problem. A genetic instruction for a spritz of testosterone in certain cells may manifest itself in several different ways upon an adult body, only one of which need be the actual target of sexual selection.

Clearly hominin females were mating preferentially with males that had other qualities than those of typical male ape. What might those criteria have been? A good reputation? Wisdom? Aesthetic sense in self-decoration, or in rock-chipping? Family connections? A sense of humor? Throwing accuracy? The most novel mating criterion of all was one that has no homolog at all in the apes: "Someone my parents will like."

Brain Differences

Whether men and women have identical brains is a question of perennial interest to sexists. A negative result is uninteresting to them, because it would suggest that, as many post-Darwinian evolutionists and creationists have acknowledged, there is no inherent difference between the cognitive processes of a random man and woman. A positive result, on the other hand, finding a real difference between the brains of men and women, might be grist for the sexist's mill, if several assumptions hold up (which, of course, they don't):

First, that there a particular relationship between thought processes and cranial features;

Second, that cranial features are the result of innate properties;

Third, that innate intellectual differences are grounds for unequal rights.

The first assumption is a staple of anecdotes about blind alleys in science. As the science of anatomy matured in the 1700s, and it became clear that the brain is the organic seat of thought, and that form and function are related, it was reasonable to suppose that big brains had big thoughts, and different-shaped brains had different-shaped thoughts. Moreover, different kinds of thoughts might be localized to different parts of the brain, which in turn could be stimulated and exercised like any muscle. Consequently, a very loving person might have an enlarged "love" part of the brain, while a math whiz might have an enlarged "math" part of the brain. This anatomical science of telling someone's personality and aptitudes from the bumps on their head came to be known as phrenology, and didn't make it into the twentieth century. While there is some localization of cerebral function, much of human thought is diffusely distributed throughout the brain; and human heads of various sizes and shapes are all capable of perfectly normal human thoughts. In other words, whatever biological differences may have existed between the Maya and the Egyptians does not appear to have involved the pyramid-building parts of the brain. Cranial biology just doesn't explain cultural achievement.

The second assumption was undermined by a major discovery of twentieth-century physical anthropology: that bodies, and especially heads, receive significant input developmentally from the external conditions of life. In a fruit fly, those conditions are largely natural – heat, light, atmosphere – but it is a lot more complicated in humans. In humans, there are irrational local aesthetic technologies, including traditions of self-mutilation and cranial deformation. Moreover, there are the effects of labor and trauma upon the body – all of which indicate that the human body is very highly adaptable, or developmentally plastic. That is to say, the range of normal human mental function embraces a diverse spectrum of human physical forms, and (aside from pathologies) the human brain seems to work just as well in all sorts of bodies and all sorts of climates.

The third assumption raises the question of why anyone even bothers to study this question. On your best sexist day, let's say that you find a blob of brain matter that is smaller or weaker or grayer in women than in men. What is the "therefore"? The size of the mastoid process in men and woman is uncontroversial; and it comes with no "therefore." But an actual physical difference of the structure of the brain would mean … what? That whatever that blob of brain matter does, it does it a bit worse in women? Obviously, those data would also be consistent with the possibility that feminine activity and masculine activity could cause the brain to grow differently. But suppose it were indeed true that there exists a heretofore unknown cerebral difference between men and women caused by the sex hormones. So what? Would that somehow show that men's and women's brains work differently? Would that explain why women shouldn't vote or earn equal wages? What would such an ana-tomical discovery mean?

This is not a case of how the data might be abused in some fashion; this is the data being collected in order to be abused. The data would in fact mean nothing, or perhaps less than nothing, if they suggested a politically evil act; the biology would simply be there as a red herring, as a spurious argument from difference to inequality.

The political movements in the US in the latter part of the twentieth century made it clear that biology is irrelevant to equality. We don't allocate rights based on your IQ. It's not that we showed in the 1960s that Blacks and Whites were actually intellectual equals (and how would we prove such a fact anyway?); it's that we showed that biological differences, and especially imaginary biological differences, are simply irrelevant to the political conversation.

Blood Differences

Male and female mammals have some blood differences. There are of course average age-related differences in the concentration of certain hormones, which can be detected in the blood. More distinct, however, are the chromosome differences detectable in the cells, as noted above. An assay for the presence of a Y chromosome will tell you that the chromosome came from a chromosomal male. An assay for an X chromosome won't be as diagnostic, since both chromosomal males and females have them. But in females, with two X chromosomes, one of them randomly condenses and goes pretty much dormant early in embryonic life. Consequently, chromosomal females, like chromosomal males, have only one functional X chromosome in each cell. Unlike in the cells of a chromosomal male, however, the condensed and minimally functional X chromosome is visible in the nuclei of female cells. (In people with XXY aneuploidy, or Klinefelter syndrome, the cells have both a Y chromosome and a condensed X chromosome.)

In other words, given what we now know about human cell biology, you can look at cells, and you can tell whether they came from a chromosomal male or female.

But that isn't what a Soviet geneticist named E. O. Manoilov was claiming in the mid-1920s. No, his claim was much simpler: Given a sample of blood, he could tell whether it came from a male or a female by adding a few reagents, shaking it up, and observing what color it turned. "The male blood soon becomes colorless or nearly so, while female blood retains its reddish-violet color." In fact, it was such a powerful assay that it didn't even require blood, because Manoilov's blood test also worked on plants. The male plants or flower parts faded in color to yellow, while the female plants or plant parts remained resolutely green. In spite of the absurdity of the claim, America's best geneticists began replicating it, and publishing on it themselves in leading scientific journals. For example, when the Manoilov blood test was given to male pigeons, they came out male, except when they were nesting, when they came out female. Why? "Because during this period the male is relatively inactive, a condition that is a feminine attribute."

With the aid of a century of hindsight, we can be impressed by how easily the geneticists bamboozled themselves by knowing what the answer was supposed to be, and then looking for it. And they worked with expectations not just for the samples that they were testing, but for the basic assumption that they brought to the work: That deep down inside, there are males and there are females and they are deep-down different, and so we should be able to detect it, as students of the deep-down inside stuff.

So too, it made perfect sense when Manoilov announced that he could distinguish the blood of different races: "For me it is absolutely clear that, by analogy to the presence of hormones characterizing this or that sex, there must be something correspondingly specific of race in the blood of different races of mankind." Except that he wasn't even distinguishing African, European and Asian bloods from one another; those weren't the groups that mattered. Manoilov had a genetic test to distinguish the blood of Russians from that of Jews. After appropriate treatment, "Jewish blood will be paler than that of Russian." And it worked in 91.7% of cases. (For those naïve enough to wonder, "What about Russian Jews?" obviously no such category existed; they were binary, discrete, and mutually exclusive to Manoilov.) It worked on other races, too. In addition to distinguishing the blood of Jews from Russians, it could soon also tell both of their bloods from Poles, Latvians, Estonians, Koreans, and Kyrgyz.

Helluva blood test. As long as you are pretty sure that the distinctions are binary and innate, this test could identify them. Anthropologist Earnest Hooton appreciated that the results were not so much wrong as impossible, for the test was finding discrete physiological distinctions among people who didn't discretely differ physiologically, but rather differed politically-nationally.

But the Manoilov blood test didn't stop there. Not only could it tell male blood from female blood, even in bloodless creatures, and tell Jewish blood from

Russian or Polish blood, but it could also detect the deep-down difference between the blood of straight and gay people. The people were deep-down different, so their bloods had to be, as well. The blood of "homosexualists" came out female, and those who "suffered from Lesbian love" came out male.

Definitely a helluva test. (In retrospect, the trick seems to be that the blood goes through a series of color changes as the procedure is followed, and all you have to do is to stop watching at the appropriate moment, in order to get the right answer.) The Manoilov blood test faded from the scientific literature in the 1930s, before Manoilov had a chance to adapt it to tell your astrological sign or shoe size. It stands, however, as a testament to the power of positive thinking in science. If you want to find a result badly enough, you can.

And it all began with telling men from women by their blood, which actually does sound reasonable enough.

Conclusions

The merits of studying constitutional differences between men and women lie in remediating inequities in health care, given their somewhat different bodies. Searching for innate differences in their intellectual or moral or emotional qualities isn't an honorable scientific pursuit. Basing claims about such differences on associations with differences in their brains or their genes is even less honorable, and is lousy science as well.

Moreover, although the categories are symbolically binary and oppositional, the people covered by them are not. Like race, sex is a symbolic category that correlates with biological difference to some extent. The binary opposition of man and woman, however, is imposed, not discovered.

8 You Are Not 2% Interestingly Exotic

Of the many things that DNA is, or can be, presumably the most reliable is as a marker of recent ancestry. You got half of your DNA from mom, and half from dad. In a situation of contested parenthood, the evidence of DNA is the gold standard. Careers in daytime television attest admirably to that fact.

With great authority, however, comes great responsibility, and responsibility isn't a scientific concept; consequently, "responsibility" represents humanists sometimes having to hold the reins on scientific practice.

The Human Genome Diversity Project

Molecular geneticists of the 1980s had a great idea: Sequence the entire human genome, all three billion nucleotides. For no particular reason, except that interesting information would presumably fall out of it, and technologically, it was beginning to look like we could. So, with the promise of using that information to cure genetic diseases, molecular geneticists got their money and genomics was born. They neglected to mention that without gene therapy, that information would take decades before it could cure anyone's genetic disease, but let's not digress.

Population geneticists of the 1990s were so impressed by the success of their medical genetics colleagues that they tried to clip off a piece of that action for themselves. Their plan was to collect information on the diversity within the human gene pool. They weren't going to cure any diseases, but they might learn something about the microevolutionary history of our species. Unfortunately, the population geneticists soon learned that nobody cared about the microevolutionary history of our species quite as much as they did, and so they tried to hop aboard the genetic disease bandwagon – different peoples have their own genetic afflictions needing cures, after all – but it was too little, too late. In 1996 they were informed that they didn't get their federal grant money.

But there was still money out there. Human genetic diversity does have subtle geographical patterns embedded, and since the 1960s geneticists had been using those patterns to try and decide which populations were more closely related to which other populations, with varying degrees of success. Moreover, every molecular geneticist worth their salt now was involved with a medical genomics start-up company of some sort. Surely there was money to be made in the private sector with the right ideas and technologies about human population genomics.

Two of the earliest ventures into private-sector human population genomics involved matching people's DNAs against one another and selling the results. African-Americans, having had their ancestral connection to African peoples largely erased by enslavement, might well be interested in seeing whose African mitochondrial DNA their own most closely resembles. Europeans might be interested to know which of seven apparent clusters their own mitochondrial DNA falls into. In 2005, National Geographic, long a leader in colonial practices, introduced the "Genographic Project" in an ambitious and creative attempt to commodify ancestry. Indigenous peoples "participated" for free (that is to say, their DNA became the comparative database), while others paid to see where in the presumptive genetic patterns of our species they fit in.

As the technology improved and the databases grew, so did the interests of venture capital, and today the big companies like 23andMe make hundreds of millions of dollars annually, purveying scientific stories of their personal ancestry to a broad clientele of paying customers (see *Understanding DNA Ancestry* by Sheldon Krimsky in this series).

An Unusual Science

Genetic ancestry testing means different things in different countries. In one place it might claim to reflect your genetic contribution from ancient peoples of the local area, based on their very subtle differences. In another place it might claim to reflect your genetic contribution from more grossly divergent or different peoples of the world. There are any number of warnings from geneticists about why and how to take the results with a grain of salt. Rather than repeat these warnings, however, we will acknowledge that genetic tests work in a fairly crude sense. Just like if you know where blonde people are found aboriginally, or where people with hooked or snub noses are found, or people with the combination of dark skin and wavy hair are found, you can make a pretty good stab at the ancestry of a person from just eyeballing them. Those traits are, broadly speaking, innate and genetically hard-wired, and thus ought to track the same patterns as the DNA variations studied by the geneticists. Both kinds of inferences about ancestry are reasonable within limits, but not particularly reliable in finer detail.

Take the case of the three identical triplets given ancestry tests for a television show. Since the Dahm triplets had blond hair, blue eyes, and narrow noses, you could guess that their ancestry was primarily from northern Europe. And their DNA concurred! But when the DNA was asked how much "French and German" ancestry the women had, the answers ranged from 11% to 22%, even though they were clones of one another. A science in which 11 and 22 are the same number is a strange science indeed, a science in which you can almost hear the statistical error bars crying out to be included.

It isn't clear how much those error bars would even help, however, for the entire project is founded on the problem of representation. After all, no one is actually analyzing your ancestry. What they are analyzing is the pattern of similarity exhibited by your DNA to the DNAs they have acquired from people around the world. In this case, "French and German ancestry" means something more like "degree of similarity to the samples that we possess, standing for the gene pool of France and Germany."

That in turn raises the question of just what it takes to be included in their sample, and to officially represent the gene pool of "France and Germany." Just how French do you have to be? Do people from Luxembourg or Switzerland or Belgium count? If not, why not? It's not as if they are so far away from France and Germany that their gene pools would never be in contact. And if so, isn't it misleading to call the sample just French and

German? Moreover, France is a big country, and the people of southern France are more similar to the people of northern Italy than to the people of northern France. That is, of course, because human genetic variation is structured geographically, not nationalistically; and while geography and nation may be correlated, they certainly do not map on to one another with any degree of precision. That is a major reason why different companies may give different results: because they are often comparing you to different sets of people ostensibly representing the same places.

Moreover, a somewhat romanticized (or perhaps a bit less charitably, racist) idea of the national gene pool underlies the enterprise. Once again, how French do you have to be, to be chosen to represent the French gene pool in a genetic study? Are you excluded if your ancestors immigrated to France from Algeria in the 1950s? And it's not as if the gene pool of France was pure before then. There were Vikings from the north in the Middle Ages, and Italians from Julius Caesar's legions in the south in Classical times, as well as many others all making (unsolicited) contributions to the French gene pool. Is the goal to "freeze" the French gene pool at some imaginary time in the past, when perhaps the people of the region were all clones of one another, and had just decided to begin nasalizing their vowels and eating snails and horses? Because that is certainly what it seems like – a very unhistorical imagining of the French gene pool.

These issues don't really matter scientifically, however, because this is not science in the way we ordinarily think of science. To begin with, this is corporate science. Nobody is saying that academic science was ever pure, but of all the conflicts of interest confronted by students of heredity and human diversity over the decades, this one – the corporate profit motive – has generally not been one that they faced. After all, when the profit motive comes into conflict with the scientific motive, which will prevail? The asbestos and tobacco industries could be consulted as depressing precedents. Or as Jesus reputedly said on a mount, "No one can serve two masters ... You cannot serve God and wealth." It was the same conflict of interest, although articulated in theological terms.

But Jesus got to the heart of the matter. Genomic ancestry tests are not a public service. They are for-profit ventures, which means that although the spread of faith in scientific research is all well and good, the spread of corporate profits is even better. And in this case, the profits stem from selling a product, and that product is a scientifically constructed story about your ancestry. Right or wrong doesn't really matter. After all, it is probably right, or close to right. Right?

And nobody is hurt if it's wrong, right? After all, this is not a genetic test to diagnose a disease, in which a health outcome may directly hinge upon the results. This is a genetic test for fun. It's recreational science. The fine print even says so.

But what does recreational science mean? That is actually rather hard to say, since it is a novelty. We don't imagine recreational cardiology or recreational nuclear physics. Certainly, there have always been amateurs, but we're not talking here about amateurs, we're talking about science, by scientists in white lab coats, running expensive machines with flashing multicolored lights. Science, but recreational science. Presumably that phrase was coined by lawyers, to disclaim any liability or legal responsibility or any uses to which the results may be put. This is science that we won't back up in court. It's all there in the fine print. You can use it for information, research, education, and recreation. *But not for anything else.*

Does that sound a bit unusual?

After all, one of the primary uses of DNA testing is specifically in a legal setting, to establish paternity or presence at a crime scene. DNA needs, and indeed has, considerable weight as evidence in our legal system. But not this particular set of DNA conclusions.

And that raises the fundamental question: If the disclaimers are structured to absolve the company of responsibility for the conclusions they sell, then why are we wasting our time arguing about their reliability? In other words, if these scientific conclusions are not reliable enough to defend in court, then they are presumably more like horoscopes (which do not carry weight in court) than like paternity analyses (which do).

To be fair, of course the ancestry tests do give you information that is sort of true. If you are naturally blond, it will validate that much of your ancestry comes from northern Europe. The big questions arise with the little numbers.

What about that 3% Korean in your blond genome? Could one of your 32 great-great-great-grandparents have been Korean? You can't account for all 32 of them, can you? Or is that 3% essentially noise in the system; it might as well be 3% Atlantean or Martian, because it simply represents genetic static?

And if you can't distinguish an actual ancestor just four or five generations ago from the statistical fog, then at least you now have a lower limit on how much of your ancestry is actually knowable in this way.

Nobody knows for sure, but I would venture to say that any single-digit number anywhere in these ancestry tests is probably indistinguishable from zero.

White supremacists were among the clients who recognized most readily that you could take the ancestry tests to validate the personal narrative of your heritage, but if it came out in a way you didn't like, there were plenty of good scientific reasons to reject the results. And they were right: clearly this is not science like anthropogenic climate change or the heliocentric solar system or the Krebs cycle or your latest MRI. This is optional science. You are purchasing a story that might well be true.

And if it isn't, who is harmed? The white supremacists? Meh.

The Downside

So your genetic ancestry actually means your genetic match not to your ancestors but to people living today, which depends crucially on two things: whose samples they are trying to compare you against, and the statistical algorithm they use to condense all of that information on your similarities to various groups into a single number. Both of those are specific to the company selling the genomic product, which is why the numbers vary from company to company.

But the accuracy of these genetic ancestry tests is presumably offset by the satisfaction on the parts of millions of consumers. Many people have discovered lost or unknown relatives, for example. Of course, that's only a good thing if we assume that those relatives want to be, or at least don't mind being, discovered – and that your connection to them is benign and not threatening. There are, after all, benign secrets and not-so-benign secrets.

Some criminal suspects have been preliminarily identified by matching a relative's DNA to the crime-scene DNA. And that's good because it helped solve a crime. But it's sort of bad too, because it does seem as though someone's privacy was invaded – notably, the relative who submitted a DNA sample for some other purpose.

The discovery of some unfamiliar ancestry might stimulate some new interest in other peoples and cultures, which is good. But on the other hand, Jeez, is that what it takes to get you interested in other peoples and cultures? They have to be somehow connected to you personally? That's pathetic. And more importantly, are you really going to start eating bratwurst or haggis because you now think you have a bit of German or Scottish DNA? Not only are those traits significantly not even genetically based, but once again, we face the genetic reification of political entities, in this case Germany and Scotland, which are not genetic categories. The 150 or so miles that separate Glasgow in Scotland from Newcastle upon Tyne in England may be marked by distinctions of dialect and government, but not by distinctions of genetics. Likewise for the 150 or so miles that separate Dusseldorf from Amsterdam. These are politically, not genetically, discrete – and we have a bit of historical knowledge concerning the consequences of confusing political categories for genetic ones.

And it's not pretty.

Let's talk about the gene pool of Germany, a large chunk of north-central Europe that didn't even become an official place until 1871. Prior to that time, north-central Europe had been composed of a loose confederation of territories, city-states, principalities, tribes, kingdoms, and burgs, inhabited by people who spoke languages similar to one another and different from Latin. Julius Caesar seems to have applied the term *Germania* to whoever lived across the Rhine from the Gauls, whom he was busy murdering. The Roman historian Tacitus wrote a book about the peoples of that area around the year 100 CE, and said that they all had red hair and blue eyes, and were basically lazy drunkards who didn't like paying taxes, and since they were illiterate, they couldn't readily respond. And that's just what we know of the Germans of 2,000 years ago.

The obvious question is: Do you have to have red hair and blue eyes to be included in the official German comparative gene pool? If you think Tacitus was wrong, what do you think is right, and why? Are its members predominantly Prussian, or predominantly Bavarian, or some mixture of residents from the different traditional states? Are there any Jews included? Time was, there used to be a significant number of them in the area.

In other words, the "German gene pool" is a myth, constructed according to someone's preconceived idea of what the gene pool of north-central Europe might once have looked like. And now the ambiguous word "population" becomes critical. First, there is the primordial, ancestral German "population," which will be somewhat different in any particular century, depending on recent patterns of gene flow, notably involving Romans (from the south), Huns (from the east), or Crusaders (from the west). So the "ancestral gene pool" of Germany is essentially timeless and Platonic, but it is effectively the *population* that you are interested in trying to reconstruct. But you don't have access to that, especially since it's imaginary anyway. What you do have access to is the gene pool of the living people of Germany, who inhabit what you have defined as the territory of interest. That *population* is real; people do live in Germany. So they will have to act as a real-world surrogate population for the ancient imaginary German population whose gene pool you are trying to reconstruct. But, of course, you can't study the gene pool of all Germans, so you have selected a sample to statistically represent them; and that statistical *population* can be analyzed and used as your comparative database.

But the population you are analyzing is only an approximation of the population that they statistically represent, which in turn is only an approximation of the population that they symbolically represent – those ancestral Germans.

There's a lot of representation going on here, ultimately falling under the umbrella of reification, or the "fallacy of misplaced concreteness." The ancestral Germans, whose DNA you may wish you had, are an abstraction. Perhaps they all looked like the actress Marlene Dietrich. Or was she unique? It's not as if there was a steady stream of Marlene Dietrichs in Hollywood. To what extent does the sample of Germans that we possess actually reflect the people that we have defined them to be a sample *of* – namely, the people of Germany? And to what extent does the target population, the people of Germany, actually reflect the people that we want them to be – namely, the people who lived in Germany a long time ago?

We do have something real, a bunch of genetic samples, which we can study and sequence and compare. But we are only pretending that we understand the complex relationship between the population of German DNA samples that we are working on, and the mythic population of timeless ancient Germans that they are standing in for. Those ancestral Germans don't even have a gene pool, except to the extent that we can imagine and invent it.

Hence the fallacy: treating something that isn't naturalistically real as if it is, essentially willing it into existence. Of course, we do that all the time, just generally not in science. It is a fact of human social existence, and we call it culture. Injustice is not a fact of nature, since justice itself is not a fact of nature, but a value. Nevertheless, we acknowledge the reality of injustice as a fact of culture, not as a fact of nature. Chimpanzees have fights and dominance hierarchies and territoriality, but they don't have systemic oppression, Christian nationalism, or Elon Musk. Humans do, but those things only have sense or meaning in the unnatural human world of money, nations, gods, and labor. But in science, we are supposed to be remaining resolutely within the domain of nature and its facts. That, alas, is clearly impossible, for science itself is conceived and performed culturally. Thus, one of the principal tasks in science studies is to identify the cultural facts that have been mistaken for natural facts, so as to try and improve the science by safeguarding against repeating the same mistake.

Suppose we compared the gene pools of the population of Houston Astros and the population of Boston Red Sox. Would we find them to be identical? Of course not. The data would be there as biological or genetic information, but they would be meaningless, except within a cultural intellectual interpretive framework. But what might any differences that we find between the gene pools of the Astros and Red Sox actually mean? Different alleles for baseball competency? Different adaptations to the climates of Texas and Massachusetts? Different numbers of players from Central America? A single Korean on the roster?

We know that people are genetically different, and we know that groups of people are genetically different. But what merits some reflection is the assumption that groups of people are units of nature, to be contrasted against one another. Some groups of people are more genetically separated than others, but none are entirely isolated, like the inhabitants of Gilligan's Island. Even the inhabitants of Pitcairn Island, initially populated by British mutineers from HMS *Bounty* and the Tahitians they kidnapped in 1789, have had genetic contact over the decades with the rest of the world. Groups of humans are not genetically discrete, and are generally established on the basis of malleable cultural criteria.

That does not, of course, prevent them from being treated as if they were comparable and natural. What about Nigerians and Swedes? You can contrast the gene pools of a population of Nigerians and Swedes as readily as you can contrast the gene pools of Astros and Red Sox. But Nigeria and Sweden are geopolitical entities, and frankly it sounds as though you are just using them as surrogate labels for "very Black people" and "very White people." Come on, admit it. Actually, though, most human groups are fluid identities and fall somewhere in between the contrast between West Africans and Scandinavians, on the one hand, and the contrast between Major League Baseball teams, on the other. Of course, we can compare and contrast them all genetically, and when we do that, we will be in a position to reify those human groups genetically – to create genetic boundaries and identities for them that do not in fact exist in nature.

One can, in fact, flip through the classic literature in human population genetics and find the casual comparison of linguistic groups (e.g., "Bantu"), nations (e.g., "France"), and tribal ethnic identities (e.g., "Karitiana") as if they were actually comparable to one another. They are comparable in a sense, as a group of genetic samples metaphorically representing the geographically separated peoples of Africa, Europe, and South America, respectively. But comparing a linguistic group to a nation and to a tribal identity is comparing apples, oranges, and raspberries – unless, that is, you equilibrate them all by imaginatively reifying them. They are biological samples, but whatever else you wish them to represent will involve assigning cultural labels to those samples. That act, however, transforms the samples metaphorically into expressions or instances of those labels. And thus the labels will seem to become naturalistically real, even though their reality actually lies in the cultural domain.

Maybe that doesn't sound like too big of a problem, but in fact it effectively naturalizes any group of people whose DNA you can collect and label, from the Hopis, through the French, the Jews, the LGBQTI, the criminals, the depressives, and the Astros lineup. In fact, the primary fallacy of human genetics in the 1950s and 1960s was that, having redefined "race" from an essential property to a gene pool (Chapter 4), the scientists believed that they had "solved the problem" of race. But pretty obviously if a race is a population, and any grouping of humans can be a population, and any population has a gene pool, then it follows that any grouping of humans can be a race, which doesn't solve anything all. That is why the "race as gene pool" was superseded by the "race as construct." It reified races as if they were just biological populations, which naturally had gene pools.

DNA ancestry testing, in other words, raises much bigger questions than "Where did you come from?"

Bigger Questions

As in all previous interactions between technology and morality, the morality lags behind the technology. Sending your DNA to a company in the hopes of learning about your ancestry is a simple and seemingly benign enough act. But of course once that information, and whatever other information your DNA has, becomes available to you, it also becomes potentially available to others: potential employers, insurers, or rivals. Sure, there are laws intended to protect your privacy, but good luck with that, for we all know that laws can be skirted or repealed, and computers and databases can be hacked.

But with DNA as a "cultural icon," your DNA easily outgrows its role as data, and often becomes almost a Ouija board to infer the presence or emergence of phenotypes that may or may not exist (yet). These inferences are based on statistical correlations between particular DNA variants and specific, often weirdly defined phenotypes, in somebody or another. Some correlations are real, some are spurious, nearly all are meaningful to somebody, and exceedingly few are causal. I've been told, from my DNA, that I have a genetic propensity to wake up at 6:40 in the morning – as if that phenotype were somehow independent of occupation, latitude, or dog ownership. But that isn't even my oddest genetic propensity. It turns out that in my 2% Neanderthal DNA, there is a genetic propensity to sneeze after eating dark chocolate. You may have to read that sentence again.

Now, obviously Neanderthals weren't sneezing after eating dark chocolate. They lived in Europe, and died out nearly 30,000 years before the cacao plant was first cultivated in Central America. But the information is there in my DNA; it's me, isn't it? And the company is just extracting the inevitable fate, the ultimate truth of my existence, from my DNA, isn't it?

That kind of thinking, though, leads to all kinds of trouble, however tempting it may be. Thousands of years of stories reinforce the lesson that consulting an omniscient oracle carries inherent dangers, for you may critically misunderstand what you are hearing. Like Oedipus or Macbeth.

Here, of course, the primary dangers lie in the likelihood of being contacted by a genetic relative you do not want to be contacted by; or in a mismatch of some sort or some degree between your DNA ancestry results and your actual personal sense of identity. For the former, we have legal disclaimers and anonymity protections (maybe); and for the latter, as the white nationalists showed, there are plenty of legitimate reasons why you can simply reject the results you receive, not the least of which is the legal liability disclaimer that you signed.

Those, however, are small potatoes compared to the questions they raise about science itself. After all, if science is supposed to be a source of authority in modern life, science hardly needs another source of erosion of that cultural authority. We've got flat-earthers, creationists, anti-vaccinators, and anthropogenic climate change deniers in the public. The idea that the scientific results of genetic testing are more or less optional hardly boosts the reputation of science in the present social climate. Rather, it opens up another front in the old "science wars," but this time the enemy is within science itself, a fifth column. After all, the results of your MRI aren't optional, nor are the results of your sickle-cell anemia genetic test, nor your paternity test. We undertake scientific tests for the public trust in their results that science has earned. Scientific tests are supposed to be different from tarot readings. But in this particular case, your scientific results are sold to you as entertaining, not authoritative. That is really weird, not just for genetics, but for all of science.

And science, as it turns out, is still recovering from the moral sensibilities of the geneticists of a century ago.

9 We Can't Breed a Better Kind of Person

Candace Owens, a right-wing media personality, told her audience in March of 2023 some pretty strange things about pit bulls and math scores. She observed that pit bulls had been bred to be fighting dogs, and quickly pivoted:

> You can breed for height, you can breed for intelligence, you can breed for stupid if you really want to … But it's important to note that just because you are bred in one way or the other, does not mean that you are locked into that permanent state. Right? So right now the most mathematically capable people in the world are Koreans. Does it mean that Koreans are forever going to be the most mathematically capable people in the world? We could start breeding Americans to do better at math. We could start doing what they do over in Korea. Or we could not, and we can remain probably the dumbest mathematical country in the world.

But neither should we remain the bioethically dumbest country in the world, one might add. Counting up her fallacies would be more of a drinking game than a scholarly exercise. In this imaginary universe, school has nothing to do with math scores; if we want to raise our country's math scores, we need to selectively breed the geeks and nerds to each other.

Sad to say, her attitude was more or less normative in the American genetics community of a century ago. Reginald C. Punnett, whose eponymous "square" is still universally taught to biology students, told his own generation of students that academic progress "is a question of breeding rather than of pedagogics; a matter of gametes, not of training." It was self-serving rhetorical bullshit back in 1905, and it still is. And the reason we know that is basically a century or so of history, both general, and particularly the history of genetics.

It is based on a fundamental misunderstanding of human biology. Specifically, it is the false assumption that genetic, innate "nature" can be opposed to cultural, learned "nurture" and that somehow human features can be ascribed to one or to the other, or even partitioned into genetic and learned components that are effectively additive. That approximates truth only rarely, and then generally only for pathological conditions. But for the range of normal features that comprise the human condition, it is simply false. Our two most fundamental evolutionary features, walking and talking, are not 50% genetic and 50% learned. They are 100% genetic and 100% learned. We humans are born with the ability to learn to speak, which is not in the DNA of an ape, and we learn to speak. Some people have disorders that compromise their ability to learn to speak, but they are rare; and some people learn to speak better or more than others, for various kinds of reasons. But for all intents and purposes, walking and talking are entirely genetic and learned.

The nature–nurture dichotomy is folk knowledge, not scientific knowledge. Prospero, in Shakespeare's *The Tempest*, calls Caliban "a born devil, on whose nature Nurture can never stick." And that is pretty much the way the early geneticist Francis Galton saw those opposing forces as well, nearly 300 years later. But of course, science advances by supplanting folk ideas about the world with more accurate ideas. The sun doesn't go around the earth, even though you can see it rise, cross the sky, and set every day. Maggots don't coalesce on dead meat from nothing, even though it may look like they do. The fundamental elements of the universe aren't earth, air, fire, and water. And nature and nurture aren't summed together to make a person, with one or the other jockeying for control of any particular feature.

Francis Galton was, however, on the cutting edge of some major scientific advances – for example, statistics. Also natural selection, the brainchild of Charles Darwin, with whom he shared a grandfather. These two interests led Galton to an alarming conclusion: Poor people were outbreeding rich people, and consequently the composition of our future population will come to consist disproportionately of the descendants of the poor.

If the poor and the rich were more or less biologically equivalent, that wouldn't be a problem, but Galton didn't think they were at all biologically equivalent. He thought that prominent characteristics, especially mental ones, were not only innate, but also different between the rich and poor. Consequently, the microevolutionary future was suddenly grim, as the lousy attributes of the poor seemed destined to swamp the genteel attributes of the wealthy. Cousin Darwin even agreed, observing at the end of *The Descent of Man* that "as Mr. Galton has remarked, if the prudent avoid marriage, whilst the reckless marry, the inferior members will tend to supplant the better members of society." What we needed was some scientific attention and legislation to address the problem. Galton called this "eugenics" in 1883 – a little helping hand for the rich, "to give the more suitable races or strains of blood a better chance of prevailing speedily over the less suitable than they otherwise would have had."

With the cachet of modern science behind it, eugenics became a vanguard of public science education globally, although with diverse national foci. In Latin America, eugenics came to mean mostly public hygiene; in England it was mostly about class differences; in the Soviet Union it was mostly about brain anatomy; but in the US, starting around 1910, it took on its most widespread and influential form – indeed, the form that most directly impressed the Germans a couple of decades later.

Eugenics in America

America in 1910 was in crisis. There had always been the indigent and poor, although they had generally been scattered in rural parts of the country. There had been some waves of poor immigrants in the nineteenth century, notably the Irish on the Atlantic coast and the Chinese on the Pacific coast. But something was different in 1910. Now there were great numbers of Italians and Jews pouring into urban slums of the northeast. They lived 20 to a room, spoke weird languages, ate weird foods, and created neighborhoods that were unsafe for a self-respecting Protestant gentleman to walk through. There was also no income tax or big federal budget, and even if there had been, the role of the government was understood to be to stimulate economic progress, not to coddle the poor.

Who could rescue American civilization from its apparent microevolutionary destiny of being overrun by immigrant hordes and their feebleminded descendants? The answer was: Gregor Mendel.

Mendel had worked with famously dichotomous traits in peas: wrinkled and round, green and yellow, tall and short, etc. And since his work had been rediscovered in 1900, it seemed as though here was the key to heredity. All differences were binary in nature. And to the leading American geneticist, Charles Davenport, human differences were likewise binary. Just as there were wrinkled and round peas, there were smart and stupid people. The stupid people had the allele for feeblemindedness, and the smart people . . . didn't.

Moreover, stupidity was the cause of poverty. Therefore, it followed that poor people had the feeblemindedness allele, which was responsible for holding them back constitutionally or innately. A big Harvard geneticist named Edward M. East explained that fully one-sixth of the American population consisted of people

> whose nervous systems are too defective for them to appreciate what is demanded of them in a modern society. This goodly quota of irresponsibles are such because of their heredity. Their children will tend to be like them. And I do not see that anything satisfactory biologically can be done about it.

Others, however, did see a solution. First, stop those irresponsibles from breeding. And second, stop others like them from entering the country. This solution came from an unlikely source: early conservation efforts. After all, if we could save the California redwood from extinction, couldn't we save the solid, intelligent American citizen?

By 1927, spearheaded by the pioneering conservationist Madison Grant, the American eugenics movement had successfully convinced Congress to restrict the immigration of Italians and Jews, as well as East Asians, and had convinced the Supreme Court that a state should be allowed to sterilize its citizens against their will. Madison Grant was himself a lawyer, not a scientist, but the leading scientists agreed and followed him, particularly the preeminent geneticist Davenport and the preeminent paleontologist Henry Fairfield Osborn, who wrote the preface to Grant's 1916 bestseller, *The Passing of the Great Race*, a classic of American racist literature. In fact, the Advisory Council of the American Eugenics Society counted almost every biologist of note in America among its ranks in 1926, from the physiologist Walter B. Cannon to the population geneticist Sewall Wright. If you took a college course in genetics in the 1920s, your textbook might very well have explained that "even under the most favorable surroundings there would still be a great many individuals who are always on the borderline of self-supporting existence and whose contribution to society is so small that the elimination of their stock would be beneficial."

Eugenics, however, was not merely the province of arcane genetic science; it was a progressive political movement that sought to improve American society scientifically. The first mailing address of the American Eugenics Society was the home of the Yale economist Irving Fisher, who predicted to a rapt audience, reported in the *New York Times* on October 16, 1929, that the stock market would go up forever. Of course, that prognostication was spectacularly falsified less than two weeks later.

Indeed, it was in large measure the Great Depression that undid the American eugenics movement, for it showed everyone that the connection between intelligence and wealth that the egghead scientists had been writing about, was tenuous at best.

Science and Power

The eugenics movement was not simply a few rogue scientists. It was mainstream science, and indeed stimulated a massively successful campaign of public outreach for its ideas. So not only was it science, but everyone knew that it was science. Indeed, to speak out against the eugenics movement was to position yourself as anti-science. You would be positioning yourself against Mendel and Darwin.

The British eugenics movement was warped by an intellectual, and ultimately personal, confrontation between biologists who thought that all heredity worked just as it did in the famous peas (the Mendelians), and those who saw gradual, continuous variation in nature, and were less satisfied with a model of discrete, binary heredity (the biometricians). The biometricians were the ones who led the British eugenics movement, which consequently pitted them against the Mendelians. Across the pond, however, Charles Davenport was establishing a zealously Mendelian approach to American eugenics. The first challenge to the American eugenicists, consequently, came from other eugenicists in England, who thought they were doing it better, because at least they weren't dumb enough to think that feeblemindedness was a simple Mendelian trait.

Davenport had established the Eugenics Record Office on Long Island in 1910, with substantial financial backing from the Carnegie Institution of Washington and the widow of railroad magnate E. H. Harriman, and he went to work on the genetics of feeblemindedness. On November 9, 1913, *The New York Times* screamed, "English Expert Attacks American Eugenic Work" as a British biometrical eugenicist named David Heron tried to warn the US about him. Heron followed it up a few weeks later with the specific large-font charges of "carelessness, inconsistency, and misinformation." Davenport took the battle to the pages of *Science*, and eventually ... nothing happened.

Now *that* is clout. When you can be called a charlatan in *The New York Times* and *Science* and not lose a step, you can be satisfied of your indestructibility. And the American geneticists continued to line up behind Charles Davenport's eugenics. Those who didn't, prudently kept their mouths shut, at least in public.

Davenport's next antagonist was the anthropologist Franz Boas, who wrote a withering critique in *The Scientific Monthly* in 1916, although without mentioning Davenport by name. But where Davenport had argued that he wanted "to improve the race by inducing young people to make a more reasonable selection of marriage mates; to fall in love intelligently," Boas invoked "the customs and habits of mankind [to] show that such an ideal is unattainable." Moreover, even the most cursory familiarity with geopolitical history demonstrated clearly that what Davenport was interpreting as an intellectual hierarchy was actually the result of cultural history, not innate qualities. "No amount of eugenic selection," Boas wrote, "will overcome those social conditions by means of which we have raised a poverty and disease-stricken proletariat, which will be reborn from even the best stock, so long as the social conditions persist that remorselessly push human beings into helpless and hopeless misery." The eugenics community managed to dismiss him largely as an anti-evolution self-interested Jew, although his colleagues at Columbia University paid him a bit more attention.

The pioneering fruit fly geneticist Thomas Hunt Morgan also worked in Schermerhorn Hall at Columbia University, and quietly began to distance himself from Davenport and the eugenicists. By the mid-1920s, as his own stature grew, Morgan even began to politely chide the eugenicists in public. But Davenport's right-hand man, Harry Laughlin, was already testifying to Congress about "the dross in America's melting pot." Laughlin demonstrated in 1922 that there was a gradient of criminality in European immigrants, with northwestern Europeans being the most law-abiding and southeastern Europeans being the least law-abiding. With such an apparent taint in their gene pools, southeastern Europeans would obviously be a terrible addition to the American gene pool. At the request of a left-leaning editor, the Johns Hopkins biologist Herbert Spencer Jennings looked carefully at Laughlin's data and realized that Laughlin's analysis had very cleverly managed to ignore his own data on the Irish, who were both very criminalistic and very northwestern European. Clearly his data were tracking levels of poverty among immigrant communities and little else. Jennings quietly withdrew from the eugenicists as well.

When the American Eugenics Society was formally incorporated in early 1926, after a few years as a less formal organization, its massive Advisory Council was a Who's Who of American genetics, science, and society. Conspicuously absent, however, was Thomas Hunt Morgan, and Herbert Spencer Jennings quickly and quietly had his name removed.

But although the biologists would not challenge the eugenics movement publicly, it received a body blow from an unexpected source. In 1925, the state of Tennessee had outlawed the teaching of evolution, and the show trial of John T. Scopes (the "Monkey Trial") held the nation's attention. Defending the community of biologists was the great civil libertarian Clarence Darrow, who took the trouble to actually read the high-school textbook out of which Scopes was charged with teaching. Darrow discovered that not only did the textbook teach students Darwinism, but it also taught them sterilization of the poor and white supremacy. And as soon as the Scopes trial was over, Darrow unloaded on the eugenicists in a popular magazine called *The American Mercury*, edited by his friend, the journalist H. L. Mencken. Darrow had no fear of the "anti-science" label, for he was, at least for the time being, the hero of American science. And he had no fear of a professional backlash, for he was a lawyer, not a biologist. "Amongst the schemes for remolding society," he wrote, "this is the most senseless and impudent that has ever been put forward by irresponsible fanatics to plague a long-suffering race." Eugenics was, to Darrow, quite literally a cult.

But no American biologist had as yet come forward to say such a thing in public. Thomas Hunt Morgan had begun to challenge Davenport's eugenic genetics, although weakly. Strong criticism had come sporadically from the social scientists, like Boas; from Catholic writers, like G. K. Chesterton, who opposed human interference with reproduction; and now from civil libertarians, who opposed government interference in family matters. But the biologists remained fearful of swimming against the tide. In 1926, Mencken himself called eugenics "mainly blather" in the pages of *The Baltimore Sun*, and the following year he published an article in his magazine *The American Mercury* by his friend Raymond Pearl, a Johns Hopkins biologist, who had been an avid eugenicist 15 years earlier. In his 1927 essay, "The biology of superiority," Pearl attacked eugenics as "propaganda" and reiterated the criticisms that had been out there for years, from the likes of David Heron and Franz Boas. But Pearl was an American biologist, and now the first to publicly break ranks with Davenport and the American Eugenics Society. That was newsworthy, and was picked up by the Associated Press, and publicized across the country. And as if to validate the biologists' timidity, Harvard abruptly ceased its recruitment of Pearl from Johns Hopkins. There were indeed serious professional risks associated with criticizing the eugenics movement.

But of course, it was all too little, too late. By the time Pearl's "The biology of superiority" appeared in print, the immigration restriction laws had been passed, and the sterilization laws had been upheld. The mainstream geneticists had won.

Then the stock market crashed, and the Germans began to take an interest in what the Americans had accomplished. In 1934, the German geneticist Eugen

Fischer celebrated his 65th birthday with a special issue of *Zeitschrift für Morphologie und Anthropologie* in his honor. The preface, by two of his students, glowed:

> We are at a turning point. For the first time in the history of the world, the *Führer* Adolf Hitler is putting into practice the insights about the biological basis of the development of peoples – race, heredity, selection. It is no coincidence that Germany is the place where this is happening: German science puts the tools in the politician's hand.

Among the contributors to the volume were both Charles Davenport and Raymond Pearl (who was no saint; he had his own crude anti-Semitic streak). Eugen Fischer himself would later be called a war criminal in the pages of *Science*. In 1936, the Nazis specifically fêted Harry Laughlin as the inspiration for their sterilization laws, with an honorary doctorate from Heidelberg University. But by then, even Davenport was mortified, and prevailed upon Laughlin to accept the honor at the German embassy, rather than travel to Europe to accept the Nazi award.

The American geneticist Hermann J. Muller took a different approach in his 1933 critique of the eugenics movement. Economic problems, he argued, needed to be solved economically, not genetically; and once we have economic parity, then we can think about improving our gene pool. He took his Marx-based ideas to the Soviet Union, and even drew the attention of Josef Stalin, but Stalin had already decided that all of genetics was Nazi balderdash, and was busy sending his geneticists off to prison. Muller barely escaped alive.

Eugenic Amnesia

Human genetics and applied eugenics, which were effectively equivalent prior to World War II, needed to be separated from one another after the war. This was accomplished in several ways. First, Charles Davenport was buried as an ancestor (he died in 1944), and a formerly obscure British medical geneticist named Archibald Garrod was installed in his place. Second, eugenics was miscast as the former province of racist cranks and loonies, not of real scientists. And third, come on, it was the Nazis. Some historians wrote critically about the American eugenics movement, but geneticists themselves moved on. Eugenics, after all, was wrong, and in science class we teach what's right.

Some scientists even tried to preach a kinder, gentler eugenics, to focus on genetic screening and non-mandatory options for families at risk for genetic disease. The word, however, was poison by the 1960s, and to survive, organizations and journals changed their names. The *Annals of Eugenics* evolved into the *Annals of Human Genetics*. The American Eugenics Society mutated into the Society for the Study of Social Biology. The goal of improving humanity genetically still resonated with some – and why not? – but improving humanity non-genetically, as social reformers since Ellen Richards (who coined the term "euthenics" in 1910 to contrast with eugenics) had argued, finally became appreciated as a more effective and humane use of resources to address poverty and urban crime.

The genetic problems of the world correspondingly shrank in scope. By 1970, the genetic problems to be addressed were actual genetic syndromes, like phenylketonuria, sickle-cell anemia, and Tay–Sachs disease, which of course affected far fewer people than feeblemindedness. Nevertheless, even the sickle-cell screening program of the early 1970s retained some of the basic problems of earlier days. In particular, there were issues about what the screening could and could not detect (unaffected heterozygotes came out the same as patients with the disease); the consequences of a positive diagnosis (Don't marry that person! If you marry, don't get pregnant! If you get pregnant, abort!); and even the overall purpose of screening. The great chemist and public scientist Linus Pauling explained in 1968:

> There should be tattooed on the forehead of every young person, a symbol showing possession of the sickle-cell gene … If this were done, two young people carrying the same seriously defective gene in single dose would recognize this situation at first sight, and would refrain from falling in love with one another. It is my opinion that legislation along this line, compulsory testing for defective genes before marriage, and some form of public or semi-public display of this possession, should be adopted.

In other words, the objective for the earliest sickle-cell screening program seemed to be to uncover and stigmatize unaffected carriers, for their hereditary imperfections. Needless to say, it didn't go over well.

Nobody's Perfect

Charles Davenport's eugenic ambition to have people "fall in love intelligently" certainly was tenacious, if ridiculous. In principle, it meant that only wealthy, White, Anglo-Saxon Protestants should have babies, for all other babies would be constitutionally inferior to theirs. In practice, it was simply a war upon the helpless. Some eugenicists tried to separate the encouragement of good baby-makers ("positive" eugenics) from the sterilization or extirpation of bad baby-makers ("negative" eugenics), but that distinction didn't really matter to policy-makers. Some eugenicists tried to introduce a distinction between the racist and xenophobic eugenics of Madison Grant, and the more presumably rigorous eugenics of the scientists. But the top scientists were sympathetic to Grant, and most of them served under him, on the Advisory Council of the American Eugenics Society.

A geneticist from MIT reviewed Madison Grant's *The Passing of The Great Race* for *Science*, the leading scientific journal in the US (and which has never deigned to review any of *my* books!). Sure, the geneticist had a few quibbles, but described it nevertheless as "interesting and valuable" and "a work of solid merit." It was especially valuable and meritorious in Germany, after it was translated in 1923. And when Nazi doctor Karl Brandt was tried for war crimes in Nuremberg, he read into the record excerpts from *The Passing of the Great Race* in his defense, to show that he was only doing what the Americans had been advocating a few decades earlier. But it didn't work; he was hanged anyway.

The primary fallacy of eugenics was built in: Once you have decided that a group of people is unfit to breed, precisely the same argument leads you to the conclusion that they are unfit to live. And it will always be easier and cheaper to kill people than to operate on them. There was actually no way to separate the racism from the science.

Second, prominence and desirability don't necessarily correspond well to one another. However much you may admire the writings of Isaac Newton or Friedrich Nietzsche, you definitely would not want to populate your town with their clones. They were quite nutty and horrid, in addition to being brilliant. Moreover, genius is singular; that's what makes it noteworthy. As H. L. Mencken wryly observed, a nation of Aristotles would be as miserable as a nation of Jesse Jameses. If all baseball players were Babe Ruths or Ty Cobbs, the sport would be worse, not better.

Third, for all the eugenics talk about posterity, tastes are notoriously fickle. An admirable quality in one time and place may not be quite so admirable a few generations down the road, when circumstances, technologies, and values may well have diverged away from what they were when the breeding program began. And who would be sitting atop the decision-making chain, declaring who shall breed? Charles Davenport? Madison Grant? Precisely the terrible people who, a century later, and judging by their effects, we might actually wish had never been born? There's irony for you!

Fourth, a century of genetics has shown that the majority of "genetic" cases of intellectual disability do not in fact run in families, but are caused by sporadic changes to the genetic material. Down syndrome is caused by an additional chromosome 21. But neither parent has Down's or is a carrier – one of them simply gave two copies of chromosome 21, instead of one, to their child in their egg or sperm, which caused the condition. It's genetic in the sense of being DNA-based, but not genetic in the sense of being familial. Fragile-X syndrome has similar discontinuities between the DNA of parent and child.

Another thing that a century of genetics has also shown is that the genome is a big place, and with three billion nucleotides, there are a lot of things that can go wrong with it. And if not actually wrong, there are a lot of ways that two genomes can be different from one another. Charles Davenport may have been smarter than a lot of people, but he was hardly God's gift to womankind. Perhaps that is why he pleaded for people to fall in love intelligently, rather than any other way. We all carry DNA variants which may be better or worse or neither, depending upon the circumstances. There is no human genome that surpasses all others; there are simply a lot of ways of being human, and fortunately rather few of them were in the genomes of the horrid Madison Grant and Charles Davenport.

Ultimately, though, eugenics wasn't just wrong, it was metaphysical. It transcended the empirical analysis of heredity. When the American geneticists, following Davenport, proclaimed that feeblemindedness was caused by a recessive allele, and thus required sterilization to get rid of it, British statistical geneticists responded by showing mathematically that it would be virtually impossible. The great majority of the rare recessive alleles are in the genotypes of asymptomatic heterozygotes, and consequently are effectively invisible. You could sterilize as many feebleminded people as you want, and yet hardly make a dent in the allele frequency. Yet they proceeded undeterred; as if it ultimately wasn't really about the allele frequency at all.

Further, while it was clear that there was little reliable knowledge about hereditary infirmity a century ago (which is why the eugenicists had to make so much of it up), nevertheless, it was also clear to anyone who cared to notice, that the ruling families of Europe had some serious hereditary problems: notably, hemophilia. The failure of the blood to clot is a terrifying physiological state, and yet by the mid-1920s it was clearly familial, and it was there in the ruling houses of Britain, Russia, Germany, and Spain. And yet nobody was talking about sterilizing them. Why not? Because their blood was just obviously better than yours. Even their hemorrhages.

And finally, "culture" had only begun to be analyzed, but it was clearly broader than "environment" or "nurture" in other species. Culture constitutes a very specific social and technological context within which any human being develops mentally. Eminence or genius or any other skill set might have an innate component, but without the historical context, interest, opportunity, and dedication, that innate component will be valueless. How perverse to imagine that it would be easier to breed a genius than to find and cultivate the geniuses that already exist! Does anyone really think that the first step towards building a teleportation device would be to clone Thomas Edison?

The only people who could think such things are those who are too heavily invested in prominence and innateness to think clearly about the human condition. But prominent, innate, and hereditary are quite different things. The eugenicists, from Galton on down, invested considerable energy in showing scientifically that prominence ran in families, and assumed that if it ran in families, it was consequently innate. And if you considered a race to be just a really, really big family, you had a ready-made theory of scientific racism.

Here, the biological interests of Francis Galton on pseudo-heredity intersected with those of his contemporary, Count Arthur de Gobineau on pseudo-history (Chapter 5). And putting them together was the brilliant/evil intellectual synthesis achieved by Madison Grant in 1916. Contradicting them, Raymond Pearl's 1927 article showed that, scientifically speaking, prominence was far more sporadic than familial. The social scientists worked to disentangle "hereditary" from "innate." After all, many things are transmitted from parent to child (hereditary) without being in the DNA (innate): such as religion, language, syphilis, and trust funds. And many things seem to be inherent and immutable without being in the DNA nucleotide sequence – such as fingerprints, handedness, weekly rhythms, birthmarks, and sexual interests.

One of the signal contributions of later twentieth-century genetics was to broaden conceptions of normalcy, and to correspondingly shrink ideas of deviation from the normal. Outside of possible unique historical conditions, there doesn't seem to be any "superior" blood type, much less a "normal" blood type. The value of being type A, B, O, or AB lies primarily in your relationship to others, and whether you can give or receive transfusions from them, rather than inhering just in your own particular antigens.

Unlike what the eugenicists thought, there are a lot of genetic ways of being normal, and actually comparatively few ways of being genetically abnormal. The treatment of genetically based disability is of unquestionable importance in the modern age, but is also a small component of disability generally, as well as being a relatively small component of overall health risks in the modern world. After all, the 3–4-year difference in average lifespan between American Blacks and Whites is due to racism, not to sickle-cell anemia.

Eugenics as Folk Heredity

Genetics today is interested in providing information and options to families at risk for inherited medical conditions. The lived interests of the family override the imaginary interests of the gene pool. We now think that people with disabilities or handicaps should be treated better, not worse. The role of government is to protect and care for its citizens. And the role of science is to make people's lives better, not shorter.

Eugenics was a sinister expression of what we can call "folk heredity" – ideas about the mysterious and wonderful relationship between parents and children that are rooted in cultural values, and are connected to scientific realities only loosely, if at all. At the time eugenics was being conceived by Francis Galton, many scientists still believed in telegony (Chapter 2) – the idea that a child can resemble a mother's earlier sexual partner, rather than her current one. It's not hard to see the moralistic argument against premarital sex there.

Of course, many scientists also believed in ranked natural groups of people – which, to the extent that group membership and rank were considered to be passed down from parent to child, is also a folk theory of heredity.

But the folk hereditary idea most glaring in eugenics is an old one called preformism (Chapter 2). The question at the cutting edge of seventeenth-century biology was: Does a baby develop and change in form within the womb, or does it just grow? Buried inside that question lay a more profound one: Do human qualities arise *de novo* every generation, or are they always there somewhere? Were you once a tiny being inside an egg of your mother, beginning to expand after mom and dad got together? And was mom a tinier being inside grandma's egg? And grandma an even tinier being in great-grandma's egg, and so on, back to Eve? The biological question was answered in a straightforward fashion: Yes, babies develop from a simple, undifferentiated fertilized egg. They weren't there infinitesimally in the ovaries of Eve (or in the testes of Adam, as the "spermists" would have it, opposed to the "ovists"). But the philosophical question remained stubbornly open. Do you really have to be carefully taught to hate and fear, as Oscar Hammerstein famously put it in *South Pacific*? Or are the hatred and fear already there inside you, just waiting to emerge when you meet someone different?

The obvious problem with the two alternatives is that, regardless of whether fear and hatred are inborn or learned, what the criteria of establishing difference are, how you apply them, to whom, and how you express it – are all very, very learned. Whether fear and hatred themselves are inborn is a trivial and scientifically antediluvian question. You learn who is different, you learn how to tell they are different, you learn what that difference means, and you learn how to deal with it.

Today there are scientists who indeed maintain that "xenophobia" is an innate universal human trait, since it is so familiar. While this may or may not be true, it ignores the last two million years or so of human evolution. We are, in a sense, born into our ancestors' prejudices, but those are the products of cultural history, not of microevolution. Over the course of the ages, it is very clear that the people who hate each other the most are not the most remote and physically distinct from one another, but simply the worst neighbors.

Today there are also people who don't want others to breed. This is more often class prejudice than germ-plasm prejudice – as if they just discovered that poor people outbreed rich people – and we already know that there is a simple solution to that particular problem. Birth control and paths to upward economic mobility, especially for women, are inducements to earn more and breed less. That is why many economically developed nations are facing sub-replacement fertility. Of course, globally the human population is still rising, but there is a big difference between the statement that "there are too many people" and the statement that "there are too many of the wrong kinds of people." The people who think there are too many people can nevertheless rest assured that the world will always be big enough for one more Harvard man.

And finally, what of the people who really want to tinker with the human genes and improve the human condition that way (instead of, say, with collective bargaining or public education or banning assault rifles as a way to improve the human condition)? Modern genetics stands in their way as well. Genes never have just a single effect, because physiology and development are complicated. An imaginary allele that increased your IQ by 10 points, which sounds great, might well increase your chances of schizophrenia or brain cancer too, which would be considerably less great. And for genes in which it is simply better to be heterozygous than to be homozygous, such as hemoglobin genes in a malarial environment, or the genes of the immune system, a goal of superior genetic uniformity would simply be disastrous.

Conclusions

We all yearn for a better world. But of the things wrong with the world, genetics is pretty far down on the list. Genetic data can help guide conservation efforts and help other species, under human guidance. But the big problems in the world are not, and have never been, genetic in nature. Genetic technologies have helped families by giving them useful knowledge about present and possible future children with genetic syndromes. But only some of those conditions are even treatable, and only a handful of people have benefited from gene therapy. Modern genetics is an augmentation to health care in the modern world, and specifically to the health of the patient and family.

What we learn from eugenics is that science is dazzling. Bad science is also dazzling, often more so than good science, and for the same reason that fake jewelry is often more dazzling than real jewelry. Sometimes the dazzle is the point.

10 Conclusions

I admit there has been a lot of negativity in this book. But that's how most scientists will tell you that science works, following the twentieth-century philosophy of Karl Popper. Science proceeds, in this view, through the falsification of hypotheses. That is the engine of science, and definitely a good thing. But scientists also generally don't like to acknowledge that their own hypotheses have been falsified. Hence the perseverance of many of the folk views of human heredity and variation, in swaths of popular culture and in shrinking niches within the scholarly community itself.

My goal here has not been to describe human variation, but to help the reader understand it. Many decades ago, a book like this one would have been a mostly descriptive survey of the peoples of the world: stocky circumpolar peoples, tall and skinny Somalis, short Indonesians, darkly complexioned Sri Lankans, pale Swedes, and so forth. Such diversity exists within our species, but as we noted in Chapter 4, the actual patterns may be a bit counterintuitive, and consequently more interesting than the features themselves. And with globalization and gene flow as the primary economic and microevolutionary forces presently operating in our species, remote peoples are no longer so remote, exotic peoples are no longer quite so exotic, and even the idea of indigeneity itself is being questioned. Moreover, fixating on other people's weird bodies no longer has the social acceptability that it once had.

Since human evolution and biology have more political valence and mythic value than other sciences, it is important to understand the nature of those cultural aspects of the science. That is what I've tried to do here, to guide the reader through a scientific minefield full of pseudoscientific mines. Knowing the genetics alone is necessary but not sufficient to make it through safely. You also need an understanding of culture and history as a guide – both in the abstract and in the specific contexts of the ostensibly genetic claims being made. Are the claims about innate behavioral differences? Between empowered and disempowered peoples? Do you really think they are finally right this time? Because they are definitely *not* apolitical.

Happily, the scientific community has been making large strides of late. In 2014, a science journalist tried to recapture the racist glamour of *The Bell Curve*, and published a book saying that the scholars who actually study human diversity are all screwed up (always a red flag from a non-specialist!). Further, the author went on, the things you thought were cultural, like the Industrial Revolution and racism, were really driven instead by microevolutionary genetics. This time, however, after an insufficiently critical published review, nearly 150 geneticists signed a letter to *The New York Times*, explaining the misrepresentation of their field in that book.

In 2022, the field of physical anthropology formally ceased to exist, as that particular scientific community consciously veered away from its racist and colonial ancestry, and toward a future as the renamed American Association of Biological Anthropologists. And in 2023, the American Society of Human Genetics published an extraordinary statement annotating the missteps of previous generations, from the advocacy of eugenics in the 1920s to its failure to publicly repudiate *The Bell Curve* in the 1990s.

Every generation of scientists is able to do things technologically that were beyond the capabilities of their advisors, and the new generation reciprocally is forbidden from doing things ethically that their advisors were able to get away with. Ideally, they grow from their predecessors' missteps. Of course, if you don't learn about your predecessors' missteps, it's very hard to learn from them. Here, the recurrent problem is a tragically consistent failure to identify research that is intellectually corrupt and empirically unreliable, particularly when it serves the interests of racism, or sexism, or genetic essentialism, or any other related and outmoded ideology. Usually the scientific community has no problem rejecting research that purports to validate the outmoded ideologies of young-earth creationism or astrology or homeopathy or ancient astronauts. And properly so. But this particular set of outmoded ideologies – the folk ideas about human diversity – has been shown time and time again to be far more harmful than others.

The harm lies in failing to recognize the question that lurks in the background of all of these presumptively scientific disputes about human variation. That question is: Why does hierarchy exist? Is it a fact of nature, of some sort, or a fact of social history, of a very different sort?

As we've seen, if it's a fact of biology, then inequality is nobody's fault, and there is no solution because there isn't really a problem. It's just Mother Nature, dictating that some people be poor and others rich, that some people be colonized and others colonizers, that some people be slaves and others masters, and that some people be workers and others owners. And we all know that it is not nice to fool Her; consequently, the unequal status quo is a good thing, and not only doesn't need to be ameliorated, but it *shouldn't* be ameliorated.

On the other hand, if hierarchy is a fact of human history ... well, then maybe we ought to do something about it. Moreover, in order to understand it, we would need to rely more on humanists – the historians – than on scientists. If hierarchies are products of history, then are they inevitable? Are they ever bad? If so, then what can be done about them? But whether we talk of hereditary aristocracy, or patriarchy, or colonialism, or other forms of exploitative social relations, at the heart is a very unscientific, moral question: Is gross inequality even acceptable? If so, then a search for its possibly naturalistic causes might well be a legitimate scientific question. But if not, then centuries of scientific nonsense about brain size, IQ score, and the like are rendered irrelevant; simply a smokescreen to try and rationalize the richer having more than the poorer, and deserving it, because of their presumptively built-in qualities.

The recruitment of science for morally disreputable goals is bad for science. What is good for science is to appreciate the moral context within which it operates, and the invariably biopolitical nature of scientific, and thereby presumably authoritative, statements about the meaning of human differences. The great Ukrainian-American geneticist Theodosius Dobzhansky laid out the distinction between equality and identity for the scholarly community at the time of the Civil Rights movement, and the distinction bears repetition. In modern society, *people are equal in spite of not being identical.*

Consequently, the patterns of their differences are irrelevant to their legal status as equals. And that means that studying human diversity has great value as a scientific project, which is why we continue to study it; but no relevance at all to a discussion of social, economic, and political inequality. Understanding the science of human diversity begins by understanding its meaning to the politics of human inequality.

Summary of Common Misunderstandings

Genetic, innate, and destined are largely synonymous. Obviously there is some overlap among the things that are biologically inherited, the things that are constitutional, and the things that are inevitable. The area of most extensive overlap is in the realm of some rare genetic diseases. What is "genetic" is simply what is transmitted biologically across generations, which, for all intents and purposes, means the chromosomes – that is to say, the functional genes and the biochemical tags associated with them. On the other hand, what is "innate" refers to certain features or impulses that may be highly predictable responses to simple stimuli, from the biochemical stimuli that trigger the development of paired limbs in an embryo, to the psychological stimuli that register embarrassment and cause blushing. And what is "destined" is, of course, the stuff of myth; for very little in biology occurs independently of its external conditions. Biological features that appear regardless of the external conditions are said to be canalized; those features that are more sensitive to their surroundings are developmentally plastic. Moreover, most genetic conditions in humans are not fully penetrant, meaning that having the allele merely gives a higher probability, not a guarantee, of the associated phenotype. While these three concepts are easy to confuse, it is helpful to appreciate that the best scientific models for understanding general human biology are simply not genetic diseases.

People are the same all over. While there are some common behavioral patterns across all people, such as a smile indicating something good and a frown indicating something bad, human societies differ in fundamental assumptions about proper behavior. Even when to suppress a smile or a frown is governed by subtle cultural rules. Not restricted to simply aesthetics or language or religion or technology or social system, culture (or at least cultural differences) represents the primary feature of human variation. Ignoring culture, we easily recognize that biologically, everyone is different. Even identical twins, who are genetically the same, may nevertheless be epigenetically significantly different. At issue, then, is not whether difference exists, but rather, the nature and patterning of those differences within our species.

Genetic variation delineates human groups. Human genetic variation is located principally within human societies, not at their boundaries; and the primary determinant of how different two people are is how far apart they (or their ancestors) once resided. While you cannot genetically tell a Dane from a Norwegian, or a Hopi from a Navajo, you may well be able to distinguish a Native American from a Scandinavian by their DNA. Of course, you can also tell them apart without looking at their DNA, since people from far apart often look different from one another. Any statement allocating genotypes to ethnic labels, however, is a probabilistic statement, not a deterministic one. You can tell a Norwegian from a Native American genetically with a high probability, and a Norwegian from a Dane genetically with a low probability. Nevertheless, since citizenship is a political status, and gene flow is an accelerating microevolutionary force in the modern world, it is not impossible to encounter a Dane with Native American genetic ancestry. But you wouldn't consider such a person any less Danish for it, would you?

Everybody has what they naturally deserve. This is an impossible proposition to falsify scientifically, since it is a moral proposition. How much, and what kind of, inequality are you willing to tolerate? Across societies? Within your own society? Whether gross inequality is acceptable to you or not, is it the result of human practices, such as slavery, colonialism, exclusionary laws, and corruption, or is it the result of those practices and some sort of important cryptic difference of nature as well? We do know about slavery and other political-economic practices as drivers of inequality. But what about genetics? Shouldn't we be looking for the genes that might also be responsible for putting the rich on top? No we shouldn't. That social history is real, but those genes are imaginary. Better to invest resources in studying the former, not the latter; and in ameliorating the inequalities that result from it.

Science is amoral. No one is beyond morality. The details may vary from place to place, but any society is based on the assumption of shared value systems. Nobody wants to learn that their neighbor is a cannibal. To assert that you are beyond good and evil is to cast moral suspicion upon yourself. To imagine that science is beyond good and evil is to invite trouble for science. Promoting the claim that science is amoral is generally just a crude rationalization for promoting science that is actually immoral. This means that scientists are obliged to confront and evaluate research on a moral axis, which they are unfortunately generally not trained for. Moreover, science is obliged not only to be good, but to be on the cutting edge of good. Science should be leading humanity morally, not simply technologically. Everyone is against evil science. You should be too.

References and Further Reading

Chapter 1

For historical and critical analysis of popular ideas about genetics and human nature in modern society, see: Bowler, P. J. (1989) *The Mendelian Revolution: The Emergence of Hereditarian Concepts in Modern Science and Society*. Baltimore, MD: Johns Hopkins University Press; Lewontin, R. C. (1991) *Biology As Ideology: The Doctrine of DNA*. New York: HarperCollins; Hubbard, R., and Wald, E. (1993) *Exploding the Gene Myth*. Boston, MA: Beacon Press; Nelkin, D., and Lindee, M. S. (1995) *The DNA Mystique: The Gene as Cultural Icon*. New York: Freeman; Griffiths, P. (2002) What is innateness? *The Monist* 85: 70–85; Meloni, M. (2016) *Political Biology: Science and Social Values in Human Heredity from Eugenics to Epigenetics*. New York: Palgrave Macmillan; Milam, E. L. (2019) *Creatures of Cain: The Hunt for Human Nature in Cold War America*. Princeton, NJ: Princeton University Press

On the classic ethnographic denial of the connection between sex and reproduction, see: Malinowski, B. (1929) *The Sexual Life of Savages in North-western Melanesia; an Ethnographic Account of Courtship, Marriage, and Family Life Among the Natives of the Trobriand Islands, British New Guinea*. New York: Harcourt Brace. For cultural inheritance as the discovery of first-generation professional anthropology, see Tylor, E. B. (1871) *Primitive Culture*. London: John Murray.

A selection of Human Genome Project metaphors: Davis, J. (1990) *Mapping the Code*. New York: Wiley; Shapiro, R. (1991) *The Human Blueprint: The Race to Unlock the Secrets of Our Genetic Script*. New York: St. Martin's Press.

Davies, K. (2002) *Cracking the Genome*. Baltimore, MD: Johns Hopkins University Press. For an early criticism of the Human Genome Project's basis in Platonic essentialism, see Walsh, B., and Marks, J. (1986) Sequencing the human genome. *Nature* 322: 590.

For nature and culture as a false dichotomy, see: Haraway, D. J. (1991) *Simians, Cyborgs, and Women: The Reinvention of Nature*. New York: Routledge; Goodman, A., Heath, D., and Lindee, M. S., eds. (2003) *Genetic Nature/Culture: Anthropology and Science Beyond the Two-Culture Divide*. Berkeley, CA: University of California Press; Keller, E. F. (2010) *The Mirage of a Space between Nature and Nurture*. Durham, NC: Duke University Press.

Some well-known but naïve sociobiological ideas about genetics and human behavior include: Wilson, E. O. (1978) *On Human Nature*. Cambridge, MA: Harvard University Press; Diamond, J. (1992) *The Third Chimpanzee*. New York: HarperCollins; Wrangham, R., and Peterson, D. (1996) *Demonic Males: Apes and the Origins of Human Violence*. Boston, MA: Houghton Mifflin.

For more critical articulations of the complexities of nature and culture in explaining features of human evolution, see: Fuentes, A. (1998) Re-evaluating primate monogamy. *American Anthropologist* 100: 890–907; Kim, N. C., and Kissel, M. (2018) *Emergent Warfare In Our Evolutionary Past*. New York: Routledge; Sterelny, K. (2012) *The Evolved Apprentice: How Evolution Made Humans Unique*. Cambridge, MA: MIT Press; Marks, J. (2015) *Tales of the Ex-Apes*. Berkeley, CA: University of California Press. For genetics and sentencing, see: Aspinwall, L. G., Brown, T. R., and Tabery, J. (2012) The double-edged sword: does biomechanism increase or decrease judges' sentencing of psychopaths? *Science* 337: 846–849.

Chapter 7

For James Watson, the misogyny in his 1968 memoir *The Double Helix* (New York, Atheneum) is now legendary. The entomologist E. O. Wilson endured a similar experience to Watson's some decades previously, following the publication of his 1975 book, *Sociobiology: The New Synthesis* (Cambridge, MA: Belknap Press), which seemed to suggest that human behavior could be understood like ant behavior, which is of course genetically hard-wired. In fact, James Watson and E. O. Wilson were erstwhile colleagues and disliked each other intensely, as they competed for resources in Harvard's biology department. See Wilson, E. (1994) *Naturalist*. New York: Island Press. Watson's 2007 memoir, *Avoid Boring People* (New York, Random House) contained disparaging comments about the intellectual abilities of Africans, which were amplified in an interview with Charlotte Hunt-Grubbe and published in the *Sunday Times* (London), eventually leading to his embarrassing departure from the UK. Watson's final interview, in which he reiterated his racist beliefs, was for a documentary called *Decoding Watson* (www.pbs.org/video/decoding-watson-ua6jjx/).

For invocations of the germ-plasm theory, see: Weismann, A. (1892) *The Germ-Plasm: A Theory of Heredity*. New York: Charles Scribner's Sons; Pearson, K. (1892) *The Grammar of Science*. London: Walter Scott; Kroeber, A. L. (1916) Inheritance by magic. *American Anthropologist* 18: 19–40; Lowie, R. (1920) August Weismann. *The Freeman* 1(11): 256.

For the inheritance of acquired characteristics, see: Anonymous (1923) Scientist tells of success where Darwin met failure. Eyes developed in newts. Demonstrates acquired qualities may be inherited. Austrian savant's laurels. Evolution would be speeded up if best characteristics could be transmitted. *New York Times* (June 3); Bowler, P. J. (1983) *The Eclipse of Darwinism*. Baltimore, MD: Johns Hopkins University Press; Krieger, N. (2005) Embodiment: a conceptual glossary for epidemiology. *Journal of Epidemiology and Community Health* 59: 350–355; Gliboff, S. (2006) The case of Paul Kammerer: evolution and experimentation in the early 20th century. *Journal of the History of Biology* 39: 525–563; Gissis, S., and Jablonka, E. (2011) *Transformations of Lamarckism: From Subtle Fluids to Molecular Biology*. Cambridge, MA: MIT Press; Gravlee, C. C. (2009) How race becomes biology: embodiment of social inequality. *American Journal of Physical Anthropology* 139: 47–57; Clarkin, P. F. (2019) The embodiment of war: growth, development, and armed conflict. *Annual Review of Anthropology* 48: 423–442.

For culture, see: Kroeber, A. L., and Kluckhohn, C. (1944) *Culture: A Critical Review of Concepts and Definitions*. Papers of the Peabody Museum, Harvard University, 47; Bonner, J. T. (1980) *The Evolution of Culture in Animals*. Princeton, NJ: Princeton University Press; Kuper, A. (2000) *Culture: The Anthropologists' Account*. Cambridge, MA: Harvard University Press; Pagel, M. (2012) *Wired for Culture: Origins of the Human Social Mind*. New York: Norton.

For human adaptability, see: Boas, F. (1912) Changes in the bodily form of descendants of immigrants. *American Anthropologist* 14: 530–562; Kaplan, B. (1954) Environment and human plasticity. *American Anthropologist* 56: 781–799; Lasker, G. (1969) Human biological adaptability. *Science* 166: 1480–1486; Bateson, P., Barker, D., Clutton-Brock, T., Deb, D., D'Udine, B., Foley, R. A., Gluckman, P., Godfrey, K., Kirkwood, T., and Lahr, M. M. (2004) Developmental plasticity and human health. *Nature* 430: 419–421.

For general overviews of the history of theories of human genetics, see: Jacob, F. (1973) *The Logic of Life*. New York: Pantheon; Müller-Wille, S., and Rheinberger, H.-G., eds. (2007) *Heredity Produced: At The Crossroads Of Biology, Politics, And Culture, 1500–1870*. Cambridge, MA: MIT Press; Müller-Wille, S., and Rheinberger, H.-G. (2012) *A Cultural History of Heredity*. Chicago, IL: University of Chicago Press.

For Point Omega, see: Teilhard de Chardin, P. (1959) *The Phenomenon of Man*. New York: Harper. For Josh Gibson, see: Ribowsky, M. (1996) *The Power and the Darkness*. New York: Simon and Schuster.

Chapter 3

Many of the ideas in this chapter are lifted from my 2002 book *What it Means to be 98% Chimpanzee* (Berkeley, CA: University of California Press) and subsequent 2009 article, What is the viewpoint of hemoglobin, and does it matter? *History and Philosophy of the Life Sciences* 31: 239–260. I really wish I didn't have to keep repeating myself.

For the prehistory of molecular evolutionary studies, see: Hussey, L. M. (1926) The blood of the primates. *American Mercury* 9: 319–321; Zuckerkandl, E. (1963) Perspectives in molecular anthropology. In *Classification and Human Evolution*, edited by S. L. Washburn. Chicago: Aldine, pp. 243–272; Simpson, G. G. (1964) Organisms and molecules in evolution. *Science* 146: 1535–1538; Dietrich, M. (1998) Paradox and persuasion: negotiating the place of molecular evolution within evolutionary biology. *Journal of the History of Biology* 31: 85–111; Suarez, E. (2007) The rhetoric of informational molecules: authority and promises in the early study of molecular evolution. *Science in Context* 20: 1–29; Sommer, M. (2008) History in the gene: negotiations between molecular and organismal anthropology. *Journal of the History of Biology*, 41:

473–528; Hagen, J. B. (2009) Descended from Darwin? George Gaylord Simpson, Morris Goodman, and primate systematics. In *Descended from Darwin: Insights into the History of Evolutionary Studies, 1900–1970*, edited by J. Cain and M. Ruse. Philadelphia, PA: American Philosophical Society, pp. 93–109.

For primate systematics and evolution, see: Fleagle, J. G. (2013) *Primate Adaptation and Evolution*. San Diego, CA: Academic Press; Tuttle, R. H. (2014) *Apes and Human Evolution*. Cambridge, MA: Harvard University Press.

For general molecular evolution and contemporary issues in organizing species scientifically, see: Graur, D. (2016) *Molecular and Genome Evolution*. Sunderland, MA: Sinauer; Bromham, L. (2016) *An Introduction to Molecular Evolution and Phylogenetics*. New York: Oxford University Press; Williams, D. M., and Ebach, M. C. (2020) *Cladistics: A Guide to Biological Classification*. New York: Cambridge University Press.

For arguing with creationists, try: Marks, J. (2021) *Why Are There Still Creationists?* Medford, MA: Polity.

Chapter 4

On the early history of the concept of race: Bendyshe, T. (1865) The history of anthropology. *Memoirs of the Anthropological Society of London* 1: 335–458; Smedley, A. (1999) "Race" and the construction of human identity. *American Anthropologist* 100: 690–702; Bethencourt, F. (2013) *Racisms: From the Crusades to the Twentieth Century*. Princeton, NJ: Princeton University Press; Bancel, N., David, T., and Thomas, D., eds. (2014) *The Invention of Race: Scientific and Popular Representations*. New York: Routledge.

On the emerging empirical complexities of human races in the twentieth century: Ripley, W. Z. (1899) *The Races of Europe*. New York: D. Appleton; Young, M. (1928) The problem of the racial significance of the blood groups. *Man* 28: 153–159, 171–176; Seligman, C. (1930) *Races of Africa*. New York: Henry Holt; Huxley, J., and Haddon, A. C. (1935) *We Europeans*. London: Jonathan Cape; Coon, C. S. (1939) *The Races of Europe*. New York: Macmillan; Montagu, A. (1942) *Man's Most Dangerous Myth: The Fallacy of Race*. New York: Columbia University Press; Caspari, R. (2009) 1918: three perspectives on race and human variation. *American Journal of Physical Anthropology* 139(1): 5–15. https://doi.org/10.1002/ajpa.20975.

On the discovery of the blood groups and their racialization: Snyder, L. (1926) Human blood groups: their inheritance and racial significance. *American Journal of Physical Anthropology* 9: 233–263; Schneider, W. H. (1995) Blood group research in Great Britain, France, and the United States between the world wars. *Yearbook of Physical Anthropology* 38(S21): 87–114; Marks, J. (1996) The legacy of serological studies in American physical anthropology. *History and Philosophy of the Life Sciences* 18: 345–362; Bangham, J. (2020) *Blood Relations: Transfusion and the Making of Human Genetics*. Chicago, IL: University of Chicago Press.

On the major features of human diversity:

(Cultural): Hallpike, C. R. (1969) Social hair. *Man* 4(2): 256–264; Ragni, G., Rado, J., and MacDermot, G. (1968) *Hair, The American Tribal Love-Rock Musical*; Geertz, C. (1973) *The Interpretation of Cultures*. New York: Basic Books.

(Developmental): Shapiro, H. L. (1939) *Migration and Environment: A Study of the Physical Characteristics of the Japanese Immigrants to Hawaii and the Effects of Environment on their Descendants*. New York: Oxford University Press; Ehrich, R. W., and Coon, C. S. (1948) Occipital flattening among the Dinarics. *American Journal of Physical Anthropology* 6(2): 181–186; Duncan, E. J., Gluckman, P. D., and Dearden, P. K. (2014) Epigenetics, plasticity, and evolution: how do we link epigenetic change to phenotype? *Journal of Experimental Zoology* 322(4): 208–220.

(Cosmopolitan/polymorphic): Lewontin, R. C. (1972) The apportionment of human diversity. *Evolutionary Biology* 6: 381–398; Barbujani, G., Magagni, A., Minch, E., and Cavalli-Sforza, L. L. (1997) An apportionment of human DNA diversity. *Proceedings of the National Academy of Sciences of the USA* 94: 4516–4519.

On clinal variation: Huxley, J. S. (1938) Clines: an auxiliary taxonomic principle. *Nature* 142: 219–220, Livingstone, F. B. (1962) On the non-existence of human races. *Current Anthropology* 3: 279–281; Handley, L. J. L., Manica, A., Goudet, J., and Balloux, F. (2007) Going the distance: human population genetics in a clinal world. *Trends in Genetics* 23(9): 432–439; Fujimura, J. H., Bolnick, D. A., Rajagopalan, R., Kaufman, J. S., Lewontin, R. C., Duster, T., Ossorio, P., and Marks, J. (2014) Clines without classes: How to make sense of human variation. *Sociological Theory* 32(3): 208–227.

On local patterns of genetic diversity: Fan, S., Hansen, M. E., Lo, Y., and Tishkoff, S. A. (2016) Going global by adapting local: a review of recent human adaptation. *Science* 354(6308): 54–59; Hift, R. J., Meissner, P. N., Corrigall, A. V., Ziman, M. R., Petersen, L. A., Meissner, D. M., Davidson, B. P., Sutherland, J., Dailey, H. A., and Kirsch, R. E. (1997) Variegate porphyria in South Africa, 1688–1996: new developments in an old disease. *South African Medical Journal* 87(6): 722–727; McKusick, V. A. (2000) Ellis-van Creveld syndrome and the Amish. *Nature Genetics* 24(3): 203–204.

For contemporary ideas about race: Fields, B., and Fields, K. (2012) *Racecraft: The Soul of Inequality in American Life*. New York: Verso. Roberts, D. (2011) *Fatal Invention: How Science, Politics, and Big Business Re-Create Race in the Twenty-First Century*. New York: New Press; Bliss, C. (2012) *Race Decoded: The Genomic Fight for Social Justice*. Stanford, CA: Stanford University Press; Benjamin, R. (2019) *Race After Technology: Abolitionist Tools for the New Jim Code*. Cambridge: Polity.

Chapter 5

For Social Darwinism, see: Sumner, W. G. (1911) *War and Other Essays*. New Haven, CT: Yale University Press, p. 177; Hofstadter, R. (1944) *Social Darwinism in American Thought*. Philadelphia, PA: University of Pennsylvania Press; Dennis, R. M. (1995) Social Darwinism, scientific racism, and the metaphysics of race. *Journal of Negro Education* 64(3): 243–252.

Darwin himself: Darwin, C. (1845) *Journal of Researches into the Natural History and Geology or the Countries Visited During the Voyage of H. M. S. Beagle Round the World, under the Command of Capt. Fitz Roy, R. N.*, 2nd edition. London: John Murray, p. 500.

On heads: Sergi, G. (1893) My new principles of the classification of the human race. *Science* 22: 290; Mitchell, P. W. (2018) The fault in his seeds: lost notes to the case of bias in Samuel George Morton's cranial race science. *PLoS Biology* 16(10): e2007008; Geller, P. L. (2020) Building nation, becoming object: the bio-politics of the Samuel G. Morton crania collection. *Historical Archaeology* 54(1): 52–70.

On intelligence genes: East, E. M. (1917) Hidden feeblemindedness. *Journal of Heredity* 8: 215–217; Chase, A. (1977) *The Legacy of Malthus: The Social Costs of the New Scientific Racism.* Urbana, IL: University of Illinois Press; Richardson, S. S. (2011) Race and IQ in the postgenomic age: the microcephaly case. *BioSocieties* 6(4): 420–446.

On mental testing: Brigham, C. (1923) *A Study of American Intelligence.* Princeton, NJ: Princeton University Press; Fish, J., ed. (2002) *Race and Intelligence: Separating Science from Myth.* New York: Lawrence Erlbaum; Bliss, R. (2023) *Rethinking Intelligence: A Radical New Understanding of Our Human Potential.* New York: HarperWave.

The synergy of hereditarian and racist science with radical politics is discussed by Tucker, W. H. (2002) *The Funding of Scientific Racism: Wickliffe Draper and the Pioneer Fund.* Urbana, IL: University of Illinois Press; Meloni, M. (2016) *Political Biology: Science and Social Values in Human Heredity from Eugenics to Epigenetics.* New York: Palgrave Macmillan; Saini, A. (2019) *Superior: The Return of Race Science.* Boston, MA: Beacon Press; Jackson, J. P., and Winston, A. S. (2021) The mythical taboo on race and intelligence. *Review of General Psychology* 25(1): 3–26.

The Bell Curve was politically influential but scientifically largely valueless. For critiques, see: Lane, C. (1994) The tainted sources of "The Bell Curve." *New York Review of Books* (1 December): 14–19; Fraser, S. (1995) *The Bell Curve Wars: Race, Intelligence, and the Future of America.* New York: Basic Books; Jacoby, R., and Glauberman, N. (1995) *The Bell Curve Debate.* New York: Times Books; Fischer, C. S., Hout, M., Jankowski, M. S., Lucas, S. R., Swidler, A., and Voss, K. (1996) *Inequality by Design: Cracking the Bell Curve Myth.* Princeton, NJ: Princeton University Press; Kincheloe, J. L., Steinberg, S. R., and Gresson, A. D. (1996) *Measured Lies: The Bell Curve Examined.* New York: St. Martin's Press. Huxley, T. H. (1894) *Evolution and Ethics and Other Essays.* London: Macmillan, p. 83.

Chapter 6

The symbolic life of primates and early hominins: Chase, P. G. (1994) On symbols and the Palaeolithic. *Current Anthropology* 35(5): 627–629. https://doi.org/10.1086/204322; Knight, C. (2008) Language co-evolved with the rule of law. *Mind and Society* 7: 109–128; Knight, C. (2010) The origins of symbolic culture. In *Homo Novus: A Human Without Illusions*, edited by U. J. Frey, C. Störmer, and K. P. Willführ. Berlin: Springer-Verlag, pp. 193–211; Chapais, B. (2008) *Primeval Kinship*. Cambridge, MA: Harvard University Press; Voorhees, B., Read, D., and Gabora, L. (2020) Identity, kinship, and the evolution of cooperation. *Current Anthropology* 61: 194–218.

Evolutionary novelties of human kinship: Rodseth, L., Wrangham, R. W., Harrigan, A. M., and Smuts, B. B. (1991) The human community as a primate society. *Current Anthropology* 32(3): 221–254; Hrdy, S. B. (2009) *Mothers and Others: The Evolutionary Origins of Mutual Understanding*. Cambridge, MA: Harvard University Press; Gettler, L. T. (2010) Direct male care and hominin evolution: why male–child interaction is more than a nice social idea. *American Anthropologist* 112: 7–21; Hawkes, K., O'Connell, J. F., Jones, N. B., Alvarez, H., and Charnov, E. L. (1998) Grandmothering, menopause, and the evolution of human life histories. *Proceedings of the National Academy of Sciences of the USA* 95: 1336–1339.

Anthropological kinship: Frazer, J. G. (1900) *The Golden Bough*, 2nd edition. London: Macmillan, quotation from Vol. I, p. 288; Fortes, M. (1983) *Rules and the Emergence of Society*. Occasional Paper #39: Royal Anthropological Institute of Great Britain and Ireland; Franklin, S., and McKinnon, S. (2001) *Relative Values: Reconfiguring Kinship Studies*. Durham, NC: Duke University Press; Zerubavel, E. (2012) *Ancestors and Relatives: Genealogy, Identity, and Community*. New York: Oxford University Press; Sahlins, M. (2013) *What Kinship Is – And Is Not*. Chicago, IL: University of Chicago Press.

Genetic ancestry and pedigree collapse: Rohde, D. L., Olson, S., and Chang, J. T. (2004) Modelling the recent common ancestry of all living humans. *Nature* 431: 562–566. Van Arsdale, A. P. (2019) Population demography, ancestry, and the biological concept of race. *Annual Review of Anthropology* 48: 227–241; Mathieson, I., and Scally, A. (2020) What is ancestry? *PLoS Genetics* 16(3): e1008624.

Chapter 7

On the human sex chromosomes: Richardson, S. S. (2013) *Sex Itself: The Search for Male and Female in the Human Genome*. Chicago, IL: University of Chicago Press.

On Darwin: The quotes are from 8: "the Negro and European ... " Vol. II, p. 388; "The chief distinction ... " Vol. II, p. 327; "To avoid enemies ... " Vol. II, p. 328; "I am aware ... " Vol. II, p. 326. See also Bryan, W. J. (1922) God and evolution. *New York Times* (February 26); Dunsworth, H. (2021) This view of wife. In *A Most Interesting Problem: What Darwin's Descent of Man Got Right and Wrong about Human Evolution*, edited by J. M. DeSilva. Princeton, NJ: Princeton University Press, pp. 183–203. The quotation from Ernst Haeckel is "so müssten Sie dieselbe geradezu zwischen den höchstentwickelten Culturmenschen einerseits und den rohesten Naturmenschen andrerseits ziehen, und letztere mit den Thieren vereinigen." Haeckel, E. 1868. *Natürliche Schöpfungsgeschichte*. Berlin: Reimer, p. 549; Fuentes, A. (2021) "The Descent of Man," 150 years on. *Science* 372: 769. https://doi.org/10.1126/science.ab j4606; also online letter by Whiten, A., Bodmer, W., Charlesworth, B., Charlesworth, D., Coyne, J., de Waal, F., Gavrilets, S., Lieberman, D., Mace, R., Migliano, A. B., Pawlowski, B., and Richerson, P. (2021).

On gender: van Anders, S. (2022) Gender/sex/ual diversity and biobehavioral research. *Psychology of Sexual Orientation and Gender Diversity*. https://doi.org/10.1037/sgd0000609; Lancaster, R., Marks, J., Fausto-Sterling, A., and Fuentes, A. (2023) Complexities of gender and sex. *Anthropology Today* 39(6): 1–2.

On ethnicity: Chandra, K. (2012) *Constructivist Theories of Ethnic Politics*. Oxford: Oxford University Press, quotation from p. 8; Brodkin, K. (1999) *How Jews Became White Folks and What That Says About Race in America*. New Brunswick, NJ: Rutgers University Press; Ignatiev, N. (1996) *How the Irish Became White*. New York: Routledge.

On evolutionary sexism: Hrdy, S. B. (1981) *The Woman That Never Evolved*. Cambridge, MA: Harvard University Press; Haraway, D. (1989) *Primate Visions: Gender, Race and Nature in the World of Modern Science*, New York: Routledge; Small, M. F. (1993) *Female Choices: Sexual Behavior of Female Primates*. Ithaca, NY: Cornell University Press; Zuk, M. (2002) *Sexual Selections: What We Can and Can't Learn about Sex from Animals*. Berkeley, CA: University of California Press. Rosenthal, G. G., and Ryan, M. J. (2022) Sexual selection and the ascent of women: mate choice research since Darwin. *Science* 375: eabi6308. https://doi.org/10.1126/science.abi6308.

On heads and brains: Geller, P. L. (2017) *The Bioarchaeology of Socio-Sexual Lives*. New York: Springer; Fine, C. (2010) *Delusions of Gender: How Our Minds,*

Society, and Neurosexism Create Difference. New York: Norton; Saini, A. (2017) *Inferior: How Science Got Women Wrong, and the New Research That's Rewriting the Story.* Boston, MA: Beacon Press. Eliot, L., Ahmed, A., Khan, H., and Patel, J. (2021) Dump the "dimorphism": comprehensive synthesis of human brain studies reveals few male–female differences beyond size. *Neuroscience and Biobehavioral Reviews* 125: 667–697. DeCasien, A. R., Guma, E., Liu, S., and Raznahan, A. (2022) Sex differences in the human brain: a roadmap for more careful analysis and interpretation of a biological reality. *Biology of Sex Differences* 13: 43. https://doi.org/10.1186/s13293-022-00448-w.

On the Manoilov blood test: "The male blood … " Satina, S., and Demerec, M. (1925) Manoilov's reaction for identification of the sexes. *Science* 62: 225–226; "Because during this period … " Anonymous (1927) Science news. Some papers presented at the Philadelphia meeting. *Science* 65: xxxv; "For me … " Manoiloff, E. O. (1927) Discernment of human races by blood: particularly of Russians from Jews. *American Journal of Physical Anthropology* 10: 11–21; Poliakowa, A. T. (1927) Manoiloff's "race" reaction and its application to the determination of paternity. *American Journal of Physical Anthropology* 10: 23–29; Naidoo, N., Štrkalj, G., and Daly, T. (2007) The alchemy of human variation: race, ethnicity and Manoiloff's blood reaction *Anthropological Review* 70: 37–43.

Chapter 8

On the Human Genome Diversity Project: Cavalli-Sforza, L. L., Wilson, A. C., Cantor, C. R., Cook-Deegan, R. M., and King, M.-C. (1991) Call for a worldwide survey of human genetic diversity: a vanishing opportunity for the Human Genome Project. *Genomics* 11: 490–491; Marks, J. (2002) "We're going to tell those people who they really are": Science and relatedness. In *Relative Values: Reconfiguring Kinship Studies*, edited by S. Franklin and S. McKinnon. Chapel Hill, NC: Duke University Press, pp. 355–383; Reardon, J. (2004) *Race to the Finish: Identity and Governance in an Age of Genomics*. Princeton, NJ: Princeton University Press; Sommer, M. (2016) *History Within: The Science, Culture, and Politics of Bones, Organisms, and Molecules*. Chicago, IL: University of Chicago Press; Radin, J. (2018) Ethics in human biology: a historical perspective on present challenges. *Annual Review of Anthropology* 47: 263–278.

On ancestry testing: Bolnick, D. A., Fullwiley, D., Duster, T., Cooper, R. S., Fujimura, J., Kahn, J., Kaufman, J., Marks, J., Morning, A., Nelson, A., Ossorio, P., Reardon, J., Reverby, S., and Tallbear, K. (2007) The science and business of genetic ancestry testing. *Science* 318: 399–400; Brown, K. V. (2018) How DNA testing botched my family's heritage, and probably yours, too. https://gizmodo.com/how-dna-testing-botched-my-familys-heritage-and-probab-1820932637; Agro, C., and Denne, L. (2019) Twins get some "mystifying" results when they put 5 DNA ancestry kits to the test. www.cbc.ca/news/science/dna-ancestry-kits-twins-marketplace-1.4980976; Phillips, A. M., and Becher, S. (2022) At-home DNA tests just aren't that reliable – and the risks may outweigh the benefits. https://theconversation.com/at-home-dna-tests-just-arent-that-reliable-and-the-risks-may-outweigh-the-benefits-194349.

On populations: Winther, R. (2022) *Our Genes: A Philosophical Perspective on Human Evolutionary Genomics*. New York: Cambridge University Press; Lummis, T. (1999) *Life and Death in Eden*. London: Victor Gollancz.

On neo-liberal genomics: Cobb, M. (2022) *As Gods: A Moral History of the Genetic Age*. New York: Basic Books.

Chapter 9

Candace Owens: www.youtube.com/watch?v=mtkm3Xva5wA, accessed 25 March 2023; Punnett, R. (1905) *Mendelism*. Cambridge: Bowes and Bowes, quotation from p. 60. Darwin, C. (1871) *The Descent of Man and Selection in Relation to Sex*. London: John Murray, quotation from Vol. II, p. 403; Galton, F. (1883) *Inquiries into the Human Faculty and Its Development*. London: J. M. Dent & Sons, quotation from p. 24.

Eugenics internationally: Stepan, N. (1991) *The Hour of Eugenics: Race, Gender, and Nation in Latin America*. Ithaca, NY: Cornell University Press; Bashford, A., and Levine, P., eds. (2010) *The Oxford Handbook of the History of Eugenics*. New York: Oxford University Press; Kühl, S. (1994) *The Nazi Connection*. New York: Oxford University Press; Rutherford, A. (2022) *Control: The Dark History and Troubling Present of Eugenics*. New York: Norton.

Eugenics in America: Kevles, D. J. (1985) *In the Name of Eugenics*. Berkeley, CA: University of California Press; Allen, G. E. (1986) The Eugenics Record Office at Cold Spring Harbor, 1910–1940. *Osiris* 2(2): 225–264; East, E. M. (1927)

Heredity and Human Affairs. New York: Charles Scribner's Sons, quotation from pp. 237–238; Spiro, J. (2009) *Defending the Master Race: Conservation, Eugenics, and the Legacy of Madison Grant*. Burlington, VT: University Press of Vermont; Sinnott, E. W., and Dunn, L. C. (1925) *Principles of Genetics*. New York: McGraw-Hill, quotation from p. 406.

Science and political power: Davenport, C. B. (1913) A reply to Dr. Heron's strictures. *Science* 38: 773–774; Spencer, H. G., and Paul, D. B. (1998) The failure of a scientific critique: David Heron, Karl Pearson and Mendelian genetics. *British Journal for the History of Science* 31: 441–452; Davenport, C. B. (1911) *Heredity in Relation to Eugenics*. New York: Henry Holt, quotation from p. 4; Boas, F. (1916) Eugenics. *Scientific Monthly* 3: 471–479; Barkan, E. (1992) *The Retreat of Scientific Racism*. New York: Cambridge University Press; Jennings, H. S. (1923) Undesirable aliens. *The Survey* 51(6): 309–312, 364; Darrow, C. (1925) The Edwardses and the Jukeses. *The American Mercury* 6: 147–157; Darrow, C. (1926) The eugenics cult. *The American Mercury* 8: 129–137, quotation from p. 137; Chesterton, G. K. (1922) *Eugenics and Other Evils*. London: Cassell; Pearl, R. (1927) The biology of superiority. *The American Mercury* 12: 257–266. Hendricks, M. (2006) Raymond Pearl's "mingled mess." *Johns Hopkins Magazine* 58(2): 50–56; Aichel, O., and Verschuer, O. v. (1934) Vorwort. *Zeitschrift für Morphologie und Anthropologie*, 34: v–vi (the original reads: Wir stehen in einer Zeitenwende. Der Führer Adolf Hitler setzt zum ersten Male in der Weltgeschichte die Erkenntnisse über die biologischen Grundlagen der Entwicklung der Völker – Rasse, Erbe, Auslese – in die Tat um. Es ist kein Zufall, daß Deutschland der ort dieses Geschehens est: Die deutsche Wissenschaft legt dem Politiker das Werkzeug in die Hand); Weidenreich, F. (1946) On Eugen Fischer. *Science* 104: 399; Muller, H. J. (1933) The dominance of economics over eugenics. *Scientific Monthly* 37: 40–47.

Eugenic amnesia: Wailoo, K., and Pemberton, S. (2006) *The Troubled Dream of Genetic Medicine: Ethnicity and Innovation in Tay-Sachs, Cystic Fibrosis, and Sickle Cell* Disease. Baltimore, MD: Johns Hopkins University Press; Marks, J. (1993) Historiography of eugenics. *American Journal of Human Genetics* 52: 650–652; Richards, E. H. (1910) *Euthenics, the Science of Controllable Environment: A Plea for Better Living Conditions as a First Step Toward Higher Human Efficiency*. Boston, MA: Whitcomb & Barrows; Pauling, L. (1968) Reflections on the new biology: foreword. *UCLA Law Review* 15: 267–272.

Nobody's perfetc: Woods, F. A. (1918) The Passing of the Great Race, 2nd edition (review). *Science*, 48: 419–420; Mencken, H. L. (1927) On eugenics. *Baltimore Sun* (May 15); Rushton, A. R. (2012) Leopold: the "Bleeder Prince" and public knowledge about hemophilia in Victorian Britain. *Journal of the History of Medicine and Allied Sciences* 67: 457–490; Graves, J. L., and Goodman, A. H. (2021) *Racism, not Race: Answers to Frequently Asked Questions*. New York: Columbia University Press; Comfort, N. (2012) *The Science of Human Perfection: How Genes Became the Heart of American Medicine*. New Haven, CT: Yale University Press; Bick, D., Bick, S. L., Dimmock, D. P., Fowler, T. A., Caulfield, M. J., and Scott, R. H. (2021) An online compendium of treatable genetic disorders. *American Journal of Medical Genetics Part C: Seminars in Medical Genetics* 187: 48–54; Jackson, C. S., Turner, D., June, M., and Miller, M. V. (2023) Facing our history: building an equitable future. *American Journal of Human Genetics* 110: 377–395; Radick, G. (2023) *Disputed Inheritance: The Battle Over Mendel and the Future of Biology*. Chicago, IL: University of Chicago Press.

Chapter 10

For the travelogue approach to human diversity, see: Wood, J. G. (1868) *The Natural History of Man*. London: George Routledge; Coon, C. S., and Hunt, E. E. (1965) *The Living Races of Man*. New York: Knopf.

On human genetics confronting its misapplications: Coop, G., *et al.* (2014) Letters: "A Troublesome Inheritance." *New York Times*, August 8. http://nyti.ms/1Hz2WC; Jackson, C. S., Turner, D., June, M., and Miller, M. V. (2023) Facing our history: building an equitable future. *American Journal of Human Genetics* 110(3): 377–395.

On equality and identity: Dobzhansky, T. (1962) Genetics and equality. *Science* 137: 112–115.

Figure and Quotation Credits

Index

Printed in the United States
by Baker & Taylor Publisher Services